Contributions to Management Science

The series *Contributions to Management Science* contains research publications in all fields of business and management science. These publications are primarily monographs and multiple author works containing new research results, and also feature selected conference-based publications are also considered. The focus of the series lies in presenting the development of latest theoretical and empirical research across different viewpoints.

This book series is indexed in Scopus.

Melita Nicotra · Marco Romano

Purpose Ecosystems

Building Collective Impact for Social and Environmental Transformation

With Contribution by Valentina Mallamaci and Silvano Zappalà

Springer

Melita Nicotra
Department of Economics and Business
University of Catania
Catania, Italy

Marco Romano
Department of Economics and Business
University of Catania
Catania, Italy

ISSN 1431-1941 ISSN 2197-716X (electronic)
Contributions to Management Science
ISBN 978-3-032-38381-5 ISBN 978-3-032-38382-2 (eBook)
https://doi.org/10.1007/978-3-032-38382-2

This work was supported by NextGenerationEU (E63C22002120006).

This Springer imprint is published by the registered company Springer Nature Switzerland AG
The registered company address is: Gewerbestrasse 11, 6330 Cham, Switzerland

To Corrado,
To Giulio,
Always.

Acknowledgments This study was funded by the European Union—NextGenerationEU, in the framework of the GRINS—Growing Resilient, Inclusive, and Sustainable Project (GRINS PE00000018—CUP E63C22002120006).

The views and opinions expressed are solely those of the authors and do not necessarily reflect those of the European Union, or can the European Union be held responsible for it.

Competing Interests The authors have no competing interests to declare that are relevant to the content of this manuscript.

Contents

About the Authors

Melita Nicotra is associate professor of Management at the Department of Economics and Business, University of Catania. She teaches Management and is a faculty member of the Ph.D. Program in Economics, Management, and Decision Making (EMADE). She also serves on the Scientific Committee of the Master in Grant Office Management. Her research focuses on the intersections of entrepreneurship, entrepreneurial ecosystems, knowledge management, digital transformation, and AI-driven organizational change. Over the years, she has built a strong national and international research profile through collaborations with leading institutions in Italy and abroad. Melita has authored numerous scientific publications in top-tier journals, as well as several monographs and book chapters with international academic publishers. She is an active reviewer for several high-ranking journals in the fields of management, innovation, and knowledge studies. Beyond her research and teaching activities, Melita has coordinated and contributed to a wide range of projects on digital innovation, ecosystem development, entrepreneurial education, and technology transfer, working in collaboration with universities, research centers, public institutions, and private organizations.

Marco Romano is full professor of Management, Digital Innovation and Transformation Management, Omnichannel Management at the Department of Economics and Business, and Director of the Master in Grant Office Management, also faculty member of the Ph.D. Program in Economics, Management, and Decision Making, University of Catania. He is an associate research fellow on technology transfer at the National Institute for Nuclear Physics, a board member of Fondazione GRINS and a member of the Sicilian advisory board of UNICREDIT SpA. He authored and co-authored numerous publications in international journals on entrepreneurial issues with respect to scientists as principal investigators, entrepreneurial universities, entrepreneurial ecosystems, technology transfer, knowledge management, exploitation strategy, business planning, and corporate governance. He serves as Associate Editor of the *Journal of Intellectual Capital*. He was Visiting Professor and Visiting Scholar at several international universities and research institutes. He held leadership positions, including Vice-President of Catania International Airport (SAC SpA),

President of the Science and Technology Park of Sicily, Board Member of the Italian Network of Scientific and Technological Parks, General Manager of the Department of Economic Development, Sicily Region, and general manager of Service Health Emergency 118 in Sicily.

List of Figures

List of Tables

List of Boxes

List of Spotlights

Chapter 1
Unpacking the Purpose Ecosystem. The Compass Framework

Across many territories a new phenomenon is emerging. Diverse actors, such as universities, social enterprises, public institutions, civic organizations, and communities, are coming together around a shared commitment to social and environmental transformation. They are not simply collaborating on projects; they are building ecosystems oriented by purpose. This book is about those ecosystems, what they are, how they work, what they need to thrive, and how their impact can be measured.

We call them purpose ecosystems. A purpose ecosystem is a multi-stakeholder configuration in which diverse actors align their activities around a genuinely shared Transformative Purpose. The purpose is not a slogan. It is not a mission statement, it is the real engine that drives decisions, allocates resources, and gives meaning to collective effort. It has a motivational dimension, since it sustains commitment, and provides a reason to collaborate beyond individual interest (George et al., 2023; Hollensbe et al., 2014). Moreover, it has a directional dimension because it orients collective decisions and prioritizes resources. And it has a relational dimension too. It is never held by a single actor alone, but emerges through the interactions of those who share it.

Without authentic purpose, there is no purpose ecosystem, only a network. An ecosystem whose purpose genuinely shapes resource allocations, governance choices, and collaborative relationships has authentic purpose.

The notion of purpose ecosystem used here does not originate as a fully developed theoretical definition, but rather as an attempt to bring coherence to practices that initially appeared only partially compatible with existing frameworks.

To make sense of purpose ecosystems, the book uses a conceptual tool called the Purpose Ecosystem Compass (PEC). A compass does not tell you where to go. It helps you find your direction. The PEC works the same way. It identifies the four fundamental dimensions that every purpose ecosystem must develop and align. First, "Enabling Conditions." It refers to the structural and relational foundations that make an ecosystem possible, including institutional quality, knowledge infrastructure, financial capacity, and the social capital that allows actors to trust and

M. Nicotra and M. Romano, *Purpose Ecosystems*, Contributions to Management Science, https://doi.org/10.1007/978-3-032-38382-2_1

cooperate. The other element is "Transformative Purpose," that is, the social and environmental direction that motivates the ecosystem and guides all its activities. The third element is "Digital Mechanisms," the specific ways in which digital technologies create value within the ecosystem, through coordination platforms, transparency tools, and learning systems. Finally, "Governance Architecture," namely the rules, processes, and accountability mechanisms through which the ecosystem makes collective decisions and remains true to its mission.

No component is sufficient on its own. Strong enabling conditions without good governance cannot protect purpose from capture by powerful interests. Sophisticated digital tools without genuine purpose just create well-connected networks without direction. The power of the PEC lies precisely in the relationships among its four components. Each chapter of this book develops one component in depth. Together, they build a coherent and comprehensive picture of purpose ecosystem functioning.

The book moves through six chapters, from foundations to applications and from concepts to instruments.

In the following, this first chapter builds theoretical foundations. It traces the evolution of ecosystem thinking from its origins in biology through the development of business ecosystems, innovation ecosystems, and entrepreneurial ecosystems, explaining what makes purpose ecosystems distinct from all of these. The chapter introduces the PEC as an integrative framework and gives particular attention to how purpose emerges within ecosystems. Purpose is not a fixed declaration that emerges from above, but it is a co-constructed, dynamic force that develops through interaction, dialogue, and shared experience. The vocabulary and conceptual architecture established here carry through the entire book.

Chapter 2 examines the enabling conditions that make purpose ecosystem emergence more or less likely in each territory. Not all territories are equally ready. Some have strong institutions, rich knowledge bases, and high levels of social capital. Others face significant gaps. The chapter develops the Purpose Ecosystem Propensity Index (PEPI), a composite indicator that measures territorial readiness by integrating structural factors (institutional quality, resource mobilization capacity, knowledge infrastructure) with relational factors such as social capital and civic participation. The chapter also develops the complementary argument that institutional intervention can support purpose ecosystem development in territories where enabling conditions are weak. However, we have to maintain the fundamental distinction between what policy can and cannot do. Policy can create the container but purpose must fill it from within.

Chapter 3 deals with Transformative Purpose. Purpose can take the form of environmental sustainability, that is, the commitment to managing natural resources responsibly, reducing greenhouse gas emissions, and contributing to ecological transitions. Moreover, purpose can take the form of social transformation, that is, related to the commitment to building communities, advancing equity, and generating shared well-being. The chapter introduces the indicators of Transformative Purpose, the output, and short-term outcome signals that allow an ecosystem to assess whether it is moving in the direction it claims to be moving.

Chapter 4 develops a functional understanding of what digital mechanisms do in purpose ecosystems. They coordinate diverse actors across organizational and geographic boundaries, create transparency about the social and environmental consequences of ecosystem activities, and support learning and adaptive management. The chapter then examines how the relationship between digital mechanisms and purpose ecosystems differs fundamentally depending on whether the ecosystem has emerged organically or has been initiated by policy. In emergent ecosystems, purpose comes first and digital tools follow. In policy-initiated ecosystems, digital infrastructure often precedes genuine collaborative relationships, creating opportunities and distinctive risks. The chapter concludes with the principle of amplification without substitution. Digital mechanisms amplify what already exists, but they cannot generate shared values and relational trust on which authentic purpose ecosystems depend.

Chapter 5 addresses governance as a component of the PEC. It develops a theory of governance for systems that are neither hierarchically controlled nor purely market coordinated. The chapter examines how governance works differently in organically emergent ecosystems (where it is informal, relational, and adaptive) and in policy-initiated ecosystems (where formal structures, accountability mechanisms, and institutional oversight are present from the outset). It then addresses how legitimacy is built without hierarchy, how power operates in systems that claim to distribute it equitably, and why the paradox of structure has no final resolution, only ongoing management.

Chapter 6 asks the fundamental evaluative question: is the compass working? It develops a unified framework for measuring impact in purpose ecosystems, arguing that genuine ecosystem impact is emergent, systemic, and territorial, not reducible to the outputs of individual actors. The chapter identifies four long-term impact dimensions. These relate to environmental sustainability transitions, social transformation, territorial development, and ecosystem institutional legacy. Therefore, Chap. 6 explains how purpose ecosystems can document their contribution to systemic change. Moreover, it examines the governance and legitimacy dimensions of impact assessment, underlining that measurement itself is a governance act and that measurement processes must be participatory, community-anchored, and oriented toward learning rather than compliance.

Box 1.1 The Purpose Ecosystem Compass (PEC)

The Purpose Ecosystem Compass is the integrative framework that structures this book. It has four components.

Enabling Conditions refer to the structural and relational foundations that make purpose ecosystem emergence possible. These include institutional quality, knowledge infrastructure, financial capacity, and the social capital that allows diverse actors to trust and collaborate.

Transformative Purpose is the social and environmental direction toward which the ecosystem orients all its activities. It encompasses social transformation such as building communities, advancing equity, generating shared well-being. At the same time, it includes environmental sustainability, such as protecting ecological systems, reducing emissions, and supporting circular economy transition.

Digital Mechanisms is specific ways in which digital technologies create value within purpose ecosystems. We refer to coordination platforms that connect diverse actors, transparency tools that make social and environmental consequences visible, and learning systems that support adaptive management.

Governance Architecture refers to the rules, processes, and accountability mechanisms through which the ecosystem makes collective decisions, maintains alignment with its Transformative Purpose, and remains accountable to the communities it serves.

The PEC does not suggest a linear process. All four components are active simultaneously and must be aligned with each other. Weakness in any one component reduces the effectiveness of all the others.

After the brief overview of the entire book and the PEC framework, as previously mentioned, this chapter develops around three main arguments. Section 1.1 traces the origins and evolution of ecosystem thinking in management, from the biological metaphor through the successive development of business, innovation, platform, entrepreneurial, and knowledge ecosystems. It identifies the core characteristics that all ecosystems share and the important differences between the major types. Section 1.2 develops a definition of the purpose ecosystem, explaining the specific function of the purpose. This is a dynamic force that shapes collective behavior and that is co-created and not imposed from above. Section 1.3 introduces the Purpose Ecosystem Compass (PEC) as the integrative framework used throughout this book, explaining its four components and how they relate to each other.

1.1 Origins and Evolution of Ecosystem Thinking

The term "ecosystem" was coined in 1930 by the British ecologist Roy Clapham to describe the physical and biological components of an environment. Arthur Tansley (1935) developed it further. He emphasized the dynamic interdependence between living organisms and their abiotic environment. In biological ecology, an ecosystem includes all living organisms in an area together with the physical environment. It is a self-regulating unit through cycles of energy, nutrients, and information (Pickett & Cadenasso, 2002; Willis, 1997). This concept, related to interdependence, emergence, coevolution, and systemic resilience, was attractive for scholars seeking to describe modern economic and innovation landscapes that are complex and networked.

In 1993, in a paper published by the Harvard Business Review, James F. Moore proposed viewing business competition not through the lens of discrete industries but of coevolving communities of organizations. Moore (1993) argued that successful companies do not merely compete within industries. They create and lead "business ecosystems" in which they coevolve capabilities. Such companies work cooperatively and competitively to support new products, satisfy customer needs, and incorporate the next round of innovation. Moore's definition was quite open. Drawing on the biological metaphor, he just assigned to the term ecosystem the capacity to capture a sense of interdependence and collective flourishing. On the one hand this conceptual openness enabled the concept to evolve across disciplinary contexts. On the other hand, it also contributed to the proliferation of partially incompatible uses that continues to challenge the field (Jacobides et al., 2018; Shipilov & Gawer, 2020).

Moore's insight was elaborated by Iansiti and Levien (2004), who developed a systematic analysis of business ecosystem roles, strategies, and health metrics. They drew on empirical analysis of the software and information technology industries and identified three archetypal ecosystem roles. Such roles are played by the keystone species, the dominators, and the niche players. At the same time, the authors developed performance metrics for ecosystem health that included productivity, robustness, and the capacity to generate niche. Their work introduced a more analytical approach to business ecosystems, moving toward a framework able to generate testable propositions.

The conceptual development of the field was characterized by two processes that Autio and Thomas (2020) identify as the primary sources of conceptual proliferation. They refer to the application of the ecosystem concept at multiple and widely varying levels of analysis, and the increasing diversity of system-level outputs that ecosystems were said to collectively produce. Both processes have generated productive theoretical diversity and conceptual confusion. Understanding them is essential for locating the purpose ecosystem concept within the broader ecosystem landscape.

With respect to levels of analysis, the ecosystem concept has been applied to phenomena ranging from the micro-organizational to the global. At one end, Adner (2006, 2017) and Jacobides et al. (2018) deploy the ecosystem concept at the level of a focal firm and its surrounding network of complementors and suppliers, emphasizing the strategic interdependencies through which a specific value proposition is assembled and delivered. At the other end, scholars working in the tradition of regional and national innovation systems have applied ecosystem thinking to describe the institutional and relational architectures that support innovation at territorial scales ranging from urban neighborhoods (Radziwon et al., 2017) to entire national economies (Carayannis & Campbell, 2009; Nambisan et al., 2019; Chesbrough et al., 2014). This multi-level quality is a great strength of ecosystem thinking. But it is also one of its principal challenges, since comparisons across levels of analysis are difficult (Autio & Thomas, 2020).

As far as the diversity of system-level outputs is concerned, the ecosystem concept has been applied to describe systems that collectively produce new products or services (Adner, 2017), business model innovations embodied in new ventures (Autio et al., 2018), pre-competitive research knowledge (Clarysse et al., 2014; Järvi et al., 2018), digital platform services (Gawer & Cusumano, 2014; Tiwana et al., 2010), regional economic development (Spigel, 2017; Stam, 2015), and, as this volume argues, systemic social and environmental transformation (Sachs et al., 2019). Each output type implies different actor configurations, different governance challenges, and different performance criteria.

Despite this diversity, the literature (synthesized most comprehensively by Autio and Thomas, 2020, and Jacobides et al., 2018) has converged on core characteristics that distinguish ecosystems from other types of organizational collectives such as supply chains, industry networks, strategic alliances, or R&D consortia. Ecosystems are characterized by high participant heterogeneity, including actors from different industries, institutional sectors, and geographic locations, whose complementary capabilities and resources make collective value creation possible. All together they produce system-level outputs that exceed what any single participant could deliver alone. Ecosystems are also characterized by structural interdependence among participants, which can take technological, economic, or cognitive forms. Finally, ecosystems rely on distinctive governance. Relationships among participants are not fully defined by formal contracts or hierarchies, but are coordinated through what Jacobides et al. (2018) call "co-alignment structures" (shared standards, architectural rules, platform governance mechanisms, and normative frameworks) that enable complementary specialization without requiring hierarchical control (Wareham et al., 2014).

Building on these shared characteristics, the literature has developed different ecosystem forms that differ primarily in terms of the nature of the focal system-level output and the governance mechanisms through which that output is produced. In literature, for example, we can find studies on business ecosystems, innovation ecosystems, platform ecosystems, entrepreneurial ecosystems, or knowledge ecosystems.

Business ecosystems represent the broadest application of the ecosystem concept in management scholarship. Moore (1993, 1996) depicted the business ecosystem as an "economic community supported by a foundation of interacting organizations and individuals" that collectively produces goods and services of value to customers. The community is formed by suppliers, leading producers, competitors, and other stakeholders who work cooperatively and competitively. In this way, they coevolve their capabilities and roles, support new products, and satisfy customer needs. Unlike in conventional industry analysis, in business ecosystem analysis the emphasis is on coevolution across industry boundaries. Value is created and owned through multiple industries and requires the coordination of diverse, complementary actors. Teece (2007) extends this multi-industry perspective and focuses on the role of complementary assets, regulatory frameworks, and institutional arrangements in shaping the competitiveness of ecosystem participants.

Innovation ecosystems, developed systematically by Adner (2006, 2017) and Adner and Kapoor (2010), focus on the alignment structure required for a specific focal value proposition to be delivered to end users. Adner's contribution was to identify the ways in which innovation outcomes depend not only on the focal innovator's technological capabilities but on the complementary innovations produced by other actors in the ecosystem. An innovator who develops a technically superior product may still fail if the complementary innovations required to realize that product's value for customers are not available. He refers to compatible components, distribution infrastructure, user education, and regulatory adaptation. Adner's

(2017) subsequent reformulation of innovation ecosystems as "alignment structures" provides a more precise and analytical concept.

Platform ecosystems have achieved particular prominence in the context of the digital economy. They are characterized by the central role of a technological platform.

Platform can be seen as a set of shared components, rules, and interfaces that enables multiple parties to interact and create value. Therefore, it can coordinate the ecosystem actors, governing access to the system-level output. The intellectual foundations lie in the economics of two-sided markets (Evans & Schmalensee, 2016; Rochet & Tirole, 2003), the architecture theory of platform design (Baldwin & Woodard, 2009; Tiwana et al., 2010), and the strategy literature on platform leadership (Gawer & Cusumano, 2002, 2008, 2014). Differently from other forms of ecosystems, in platform ecosystems, we can distinguish and observe the role of the platform owner or sponsor, who controls the technological architecture, governance rules, and access conditions of the ecosystem. Moreover, the network effects are relevant in generating ecosystem value. As more users and complementors join, the value of participation increases for all participants. They create positive feedback dynamics that can lead to rapid ecosystem growth. Finally, the governance of platform ecosystems has become a central theme in digital strategy research (Gawer & Phillips, 2013; Parker et al., 2016; Nambisan et al., 2019; Wareham et al., 2014).

The study of entrepreneurial ecosystems explicitly focuses on the territorial and institutional conditions that support new venture creation as the primary system-level output. The entrepreneurial ecosystems literature has been shaped by the intersection of regional economics, entrepreneurship research, and economic geography, with important contributions from practitioners and policy communities as well as academics. In 1993, Van de Ven identified the industrial infrastructure facilitating entrepreneurship as a multi-actor, institutionally embedded system. Later contributions came from Neck et al. (2004) and Cohen (2006), who developed early frameworks for sustainable entrepreneurial ecosystems, and from Isenberg (2010, 2011), whose influential practitioner-oriented model organized around six domains brought the concept into wide policy use. Spigel (2017) provided the most analytically rigorous account of entrepreneurial ecosystem attributes, distinguishing cultural, social, and material dimensions of ecosystem support.

Recent contributions (see Corrente et al., 2019; Nicotra, 2019) have extended these frameworks by applying empirical and computational methods to map ecosystem dynamics and benchmark territories.

The *entrepreneurial ecosystem* literature shares with the broader ecosystem tradition the conviction that new venture creation is fundamentally a systemic phenomenon (Audretsch et al., 2019; Bejjani et al. 2023; Cantner et al. 2021; Cunningham et al. 2019). As Nicotra et al. (2018a) demonstrate in their systematic analysis of the causal relations between entrepreneurial ecosystem factors and entrepreneurial outcomes, the entrepreneurial ecosystem is best understood as a set of interdependent factors coordinated in a way that enables the accumulation of various forms of capital needed to feed productive entrepreneurship. Financial capital is related to funding mechanisms that provide early-stage firms with the resources needed to survive and grow (Levie & Autio, 2008). Besides, knowledge capital is the stock of know-how, human talent, research competence, and entrepreneurial education present in a territory (Bahrami & Evans, 1995; Florida 2002; Zucker et al., 1998). Institutional capital includes the formal and informal rules, regulatory frameworks, and organizational support structures (including incubators, accelerators, science parks, and legal and professional service networks) that reduce the costs and risks of entrepreneurial activity (Baumol, 1990, 1993; North, 1990; Romano et al., 2014a, 2014b). Social capital, finally, is the networks, norms, and trust relationships that facilitate the exchange of information, resources, and opportunities among ecosystem actors, including role models, mentors, and the culture of entrepreneurship embedded in a territory (Bourdieu, 1986; Coleman, 1990; Putnam, 1993).

The Nicotra et al. framework also provides a systematic approach to measuring eco-factors and eco-output and allowing rigorous causal inference, addressing a fundamental challenge in entrepreneurial ecosystem literature. As Stam (2015) and Alvedalen and Boschma (2017) have noted, the existing literature has largely relied on rich case descriptions of successful ecosystems (Silicon Valley as analyzed by Saxenian, 1994; Boulder, Colorado as analyzed by Neck et al., 2004, and Feld, 2012; and various other innovation clusters) that are valuable as existence proofs but insufficient as a basis for generalizable causal claims. By synthesizing the available indicator frameworks from global data sources and linking them to a theoretically grounded account of the mechanisms through which eco-factors generate eco-output, Nicotra et al. (2018a) provide the foundational building blocks for an empirically rigorous ecosystem science.

Knowledge ecosystems have emerged more recently in literature, driven by the growing importance assigned to collaborative knowledge creation as a driver of innovation and economic development. In knowledge ecosystems, the primary system-level output is new knowledge, collectively generated by universities, public research institutions, firms, and government bodies (Clarysse et al., 2014; Järvi et al., 2018; Nicotra & Romano, 2018; Valkokari, 2015). The knowledge ecosystem concept is distinguished from the innovation system literature by its emphasis on ecosystem-level dynamics. We refer to the co-evolutionary adaptation of actors around shared research themes, the role of co-alignment structures in coordinating collaborative knowledge production, and the distinctive governance challenges of managing pre-competitive knowledge as a shared resource. Järvi et al. (2018) define knowledge ecosystems as configurations in which participants engage in "collaborative exploration of new knowledge as activity and output central to the ecosystem," emphasizing the process-oriented, exploratory character of knowledge ecosystem activity. In this context, the absorptive capacity of territorial actors is a crucial determinant of how effectively knowledge ecosystem dynamics lead to innovation and broader societal value (Nicotra & Romano, 2018; Nicotra et al., 2014).

Across all of these ecosystem types, a set of cross-cutting dynamic processes shapes ecosystem evolution over time: emergence, competition, coevolution, and resilience. These processes describe how purpose ecosystems develop, sustain, and transform themselves.

Ecosystem *emergence* is the process through which a set of actors gradually connects and develops into a structured, coherent community able to generate collective system-level outputs. This may occur in two ways. First, through organic, bottom-up self-organization driven by the discovery of complementarities among actors. Second, through deliberate, orchestrated efforts by a lead firm, government body, or civic organization (Dattée et al., 2018; Hannah & Eisenhardt, 2018; Moore, 1993). This distinction implies different governance mechanisms, different roles for ecosystem leaders, and different vulnerabilities to ecosystem failure. Spontaneously emerged ecosystems may be more resilient and adaptive because they have developed their coordination mechanisms through evolutionary selection. However, they may also be more fragile in facing deliberate disruption or rapid technological change. Differently, orchestrated ecosystems may be more efficiently designed and strategically focused, but they risk over-dependence on the orchestrating actor.

Ecosystem *competition* operates at multiple levels simultaneously. Within ecosystems, actors compete for roles, resources, and the distribution of value. This form of competition is structurally different from conventional market competition because it is embedded in, and constrained by, the cooperative relationships that make ecosystem value creation possible (Brandenburger & Nalebuff, 1996; Jacobides et al., 2018). Between ecosystems, competition operates among entire systems for users, complementors, investment, and the legitimacy that attracts further participants (Cennamo & Santaló, 2013; Cennamo, 2019). At the territorial level, competition between ecosystems shapes the relative performance of regional economies. Here ecosystems compete for talent, capital, and innovative firms. This has become a central concern of innovation policy (Stam, 2015; Spigel, 2017). Generally, in competitive dynamics, a trade-off exists between maintaining competitive pressure to drive innovation and efficiency and, at the same time, preserving the cooperative relationships to enable collective value creation.

Coevolution describes the mutual adaptation processes through which ecosystem participants and their environment simultaneously shape and are shaped by others over time. Moore (1993) introduced the concept to capture the dynamic, non-linear character of ecosystem development. The strategic choices of a single actor generate environmental changes that in turn constrain and enable the choices of all other actors. Coevolution refers to the mutual adaptation of technologies and business models (Tiwana et al., 2010), to the co-development of institutional frameworks and organizational practices (Lewin & Volberda, 1999), and to the iterative adjustment of governance structures in response to changing ecosystem conditions (Wareham et al., 2014). The co-evolutionary perspective has important implications for ecosystem governance. Effective governance must be adaptive and iterative rather than static. It should continuously respond to the emergent dynamics of actor interaction rather than impose a fixed coordination logic.

Ecosystem *resilience* is the capacity of an ecosystem to absorb disturbances, to maintain its core functions, and to adapt to new conditions without losing its fundamental identity. It has emerged as a critical concern in the academic and policy literatures, particularly in relation to technological change, geopolitical disruption, and environmental transition (Iansiti & Levien, 2004; Wareham et al., 2014). Drawing on the biological concept of ecological resilience developed by Holling (1973), scholars have identified several structural and governance conditions that contribute to ecosystem resilience, that is, diversity of actors and capabilities, redundancy of key functions, modularity of ecosystem architecture that limits the propagation of local failures, and adaptive governance mechanisms that enable rapid collective response to external shocks. In purpose ecosystems, resilience is not only the capacity to maintain system functions under stress, but the capacity to continue generating social and environmental value.

This review of ecosystem types and dynamics (see Table 1.1) provides the conceptual foundation for understanding where the purpose ecosystem sits within the broader ecosystem landscape and what is novel about it as a theoretical construct. Existing literature has been mainly concerned with ecosystems that produce economic value. It has not adequately theorized the production of social and environmental value as a system-level output. Scholars have focused on the strategies and economic interests of profit-oriented ecosystem actors. The role of communities, civil society organizations, and public institutions as ecosystem actors with distinct logics and value orientations has been overlooked. Therefore, governance has been treated as a mechanism for coordinating economic value creation and appropriation and not for advancing equity, inclusion, and environmental sustainability. These limitations define the theoretical space that the purpose ecosystem concept and the PEC are designed to fill.

Table 1.1 Ecosystem types

Ecosystem type	Output	Key actors	Main governance challenge
Business ecosystem	Products and services across industries	Firms, suppliers, complementors	Balancing competition and cooperation
Innovation ecosystem	Aligned value propositions	Focal innovator, complementors, users	Managing alignment across interdependencies
Platform ecosystem	Transactional services via shared platform	Platform owner, users, developers	Balancing openness and control; avoiding monopoly
Entrepreneurial ecosystem	New venture creation	Firms, universities, investors, government	Building and maintaining capital stocks
Knowledge ecosystem	Pre-competitive knowledge	Universities, research centers, firms	Managing shared knowledge as a common resource
Purpose ecosystem	Social and environmental transformation	All of the above plus communities and civil society	Maintaining purpose alignment, ensuring equity, and legitimacy

Source Authors' own elaboration

1.2 Where Purpose Emerges: Co-Construction, Shared Mission, and Transformative Collaboration

The notion of purpose addresses the question of an organization's reason for being and the beneficiaries of its action (George et al., 2023; Hollensbe et al., 2014). The answer is not simply given by a formal charter or mission statement. It needs to be negotiated and co-constructed among the multiple actors who have stakes in the organization's activities. For single organizations, this complexity is challenging enough. For multi-actor ecosystems it is greater because they need to articulate and sustain a shared purpose that spans organizational, sectoral, and institutional boundaries.

In recent decades, as social and environmental pressures have intensified, a growing number of scholars and practitioners have begun to question whether profit maximization alone can adequately describe or guide how firms behave (Gartenberg & Yiu, 2023; Mayer, 2021). This dissatisfaction has brought corporate purpose to the center of the conversation. A rich body of work is now emerging, drawing on foundational contributions from stakeholder theory (Freeman, 1984), institutional theory (DiMaggio & Powell, 1983; Scott, 1995), and the resource-based view (Barney, 1991; Teece et al., 1997). It argues that sustainable organizational performance requires alignment around a purpose that extends beyond profit to encompass the creation of value for employees, communities, customers, and the natural environment (Elkington, 1997; Mackey & Sisodia, 2013; Nicotra, 2014; Porter & Kramer, 2011). Table 1.2 synthesizes the principal contributions, moving from early organizational definitions toward more collective and territorial understandings.

Despite growing interest, purpose remains theoretically underdeveloped (Morrison & Mota, 2021). First, the mechanisms through which purpose influences behavior and performance outcomes are poorly understood. This is a particularly important aspect in multi-actor settings where purpose must be shared and maintained across heterogeneous organizations with different interests and value systems. Moreover, the problem of purpose-washing remains inadequately theorized. We should ask and try to answer if purpose is substantive or rhetorical (Crane et al., 2014).

We try to address these gaps by shifting the unit of analysis from the individual organization to the ecosystem, and by developing a dynamic, emergent understanding of purpose as a collective phenomenon, that is, continuously co-constructed rather than unilaterally declared.

In this volume, a purpose ecosystem is understood as a multi-stakeholder system in which diverse actors come together around a shared societal purpose, collaborating to create social value and to generate measurable impact (Brown & Jones, 2021; Emerson & Nabatchi, 2015).

Purpose ecosystem is multi-stakeholder. It includes actors from multiple organizational types and institutional sectors; firms, governments, universities, nonprofit organizations, community groups, social enterprises, and individual citizens, each with distinct organizational logics, resource endowments, and accountability relationships. This characteristic is a constitutive requirement, because the social and environmental challenges that purpose ecosystems address can only be tackled through

Table 1.2 Definitions of purpose in the literature

Author(s)	Definition	Relevance for purpose ecosystems
Bartlett & Ghoshal (1994)	"The statement of a company's moral response to its broadly defined responsibilities, not an amoral plan for exploiting commercial opportunity"	First explicit framing of purpose as ethical and relational.
Hollensbe et al. (2014)	Organizations with purpose connect their activities "to larger values that promote the Well-being of society and individuals within and outside of business"	Shifts purpose from firm-level identity to societal orientation.
Henderson & Van den Steen (2015)	"A concrete goal or objective for the firm that reaches beyond profit maximization"	Operationalizes purpose as measurable direction.
Hurth et al. (2018)	"An organization's meaningful and enduring reason to exist that aligns with long-term financial performance, provides a clear context for daily decision-making, and unifies and motivates relevant stakeholders"	Introduces the enduring, motivational, and unifying character of purpose.
Mayer (2021)	"Finding ways of solving problems profitably where profits are defined net of the costs of avoiding and remedying problems"	Reframes the relationship between purpose and economic value.
George et al. (2023)	Purpose as "the organization's reason of being and the beneficiaries of its action," which "needs to be negotiated and co-constructed among the multiple actors who have stakes in the organization's activities"	Moves purpose from leadership declaration to co-construction across multiple stakeholders.
Kania & Kramer (2011)	"The commitment of a group of actors across multiple sectors and the community to a common agenda for solving a complex social problem"	First definition operating explicitly at the collective, cross-sector-level.
Mazzucato (2018)	Missions as "bold, inspirational" directional goals "with wide societal relevance" that set the direction for diverse public and private actors without prescribing the solution pathways	Extends purpose to the policy and territorial level.
Mulgan (2019)	Social innovation as "innovations that are social both in their ends and in their means" purposeful change that transforms not only outcomes but the relationships and governance structures through which change is produced	Introduces the systemic and relational dimension of purpose.

Source Authors' own elaboration

the coordinated action of actors whose complementary capabilities and legitimacies span the boundaries of any single organizational type (Adner, 2017; Jacobides et al., 2018; Kania & Kramer, 2011).

Besides, *purpose ecosystem is oriented toward a shared societal purpose.* This distinguishes it from other ecosystem types in terms of the nature of their system-level output. Business ecosystems produce economic value. Platform ecosystems produce transactional services. Entrepreneurial ecosystems produce new ventures. Purpose ecosystems are oriented toward societal transformation, reducing inequality, advancing environmental sustainability, expanding access to education and health, or addressing other dimensions of social and environmental challenge that markets and governments, acting alone, have proven insufficient to address. Crucially, as we argue below, this shared societal purpose is never fully defined in advance. It emerges and takes shape through the collaborative process itself.

Actually, *the ecosystem is built around co-creation.* The generation of social value in purpose ecosystems is not a product of any single actor's activities. It emerges from the interactions, relationships, and collaborative processes of diverse actors. This co-creative character implies that assessing a purpose ecosystem's performance requires attending to the quality and inclusiveness of its collaboration processes, not only to the outputs and outcomes of individual activities. Collaborative practices for co-creation, hybrid learning, participatory decision-making, and community engagement are themselves constitutive of the ecosystem's identity, because the quality and inclusiveness of collaboration determines whether the stated commitment to shared purpose is substantive or rhetorical.

Purpose ecosystems reach beyond the individual goals of any single participating actor, directing their energy toward social, environmental, and broader societal outcomes (Sachs et al., 2019). This introduces a commitment to certain values and outcomes that conventional governance frameworks are not designed to handle. Purpose ecosystems must not only coordinate diverse actors around a shared strategy. They must also maintain the alignment of those activities with the ecosystem's social and environmental mission in the face of persistent pressures that tend to erode purpose commitments over time. These governance challenges are examined in depth in Chap. 5.

Purpose is not deliberated in advance. It is not written in a founding document, decided by a steering committee, or assigned by a funding body. Actors who share a concern about a social or environmental condition, a frustration with existing institutional responses, or a vision of a different future begin to collaborate around that shared horizon. Over time, through dialogue, joint action, and collective learning, what they care about becomes recognizable as a shared purpose. Purpose is not declared. It is discovered.

The concept of purpose as emergent is consistent with the theory of complex adaptive systems (Axelrod & Cohen, 1999; Holland, 1992; Stacey, 1995). In such a theory, macroscopic order arises through the aggregation and interaction of individual actions rather than through the imposition of order from above. Ecosystem participants do not implement a purpose that has been defined for them. Through their participation in activities, governance processes, and community dialogues, actors continuously shape and redefine that purpose. It becomes visible and real when diverse actors interact over time within structures designed to support genuine collaboration. It crystallizes gradually as stakeholders discover common ground and converge around what genuinely matters for their community and territory. It functions not as a fixed destination but as what we might call an evolutionary attractor. It is a shared horizon that orients collective action without prescribing it. It shapes decisions but remains open to reinterpretation as circumstances change and the ecosystem learns.

This has several important implications for ecosystem design and governance. One aspect is that it is difficult to design a purpose ecosystem from the top-down. What can be designed are the conditions that make purpose emergence possible. This is related to relational spaces, flexible governance arrangements, shared infrastructure, and cultural norms that allow diverse actors to interact genuinely, build trust, and discover what they share. Another point concerns the fact that formal governance structures do not create purpose. At their best, they recognize it, make it visible, and create conditions for it to develop and deepen over time. They also protect purpose from being captured by the interests of the most powerful actors. Actually, purpose is inherently dynamic. It evolves in relation to new challenges, new actors, and new knowledge. A governance architecture that tries to fix purpose permanently will eventually constrain rather than enable the ecosystem's transformative potential.

Purpose ecosystems flourish where control is lost. When governance is too rigid, prescribing priorities, procedures, and indicators from above, it tends to suppress the very creativity and adaptability that purpose ecosystems depend on. The ecosystems that thrive are those that adopt more flexible governance with structures that evolve, integrate new actors, respond to emerging technologies, and leave room for experimentation and learning. This does not mean that governance is unimportant. On the contrary, the emergent quality of purpose makes governance design more important, not less.

It is worth reflecting on a significant and growing trend in contemporary innovation policy: the deliberate construction of ecosystems through top-down institutional mechanisms. Instruments such as the Italian National Recovery and Resilience Plan (PNRR), that funded national innovation ecosystems in Italy, represent substantial public investments explicitly aimed at building ecosystem structures at the territorial scale. These policy initiatives play a genuinely important and valuable role. They mobilize financial resources that individual actors could never assemble independently. They create institutional legitimacy for multi-actor collaboration that might otherwise struggle to gain recognition. They lower the transaction costs of connecting actors who share values and concerns but lack the organizational infrastructure to work together. They build the enabling conditions (infrastructure, regulatory frameworks, and financial mechanisms) that Chap. 2 of this volume measures through the Purpose Ecosystem Propensity Index.

What policy initiatives of this kind create, however, is the container rather than the content. They establish structural preconditions for purpose ecosystem development. What they cannot generate is the purpose itself. Purpose requires actors who genuinely share values, concerns, and a vision of the change they want to produce together. When those shared values are already present in a territory, policy instruments can activate and amplify them, accelerating the emergence of ecosystems that might otherwise have taken longer to form or lacked the resources to sustain themselves. When those shared values are not yet present, or when the design process bypasses the communities, whose participation is essential, the result is likely to be a formally correct structure (with governance documents, designated actors, and reporting requirements) but without the authentic shared commitment that distinguishes a genuine purpose ecosystem from a funded network.

This distinction between container and content is not a critique of top-down policy approaches. It is a design principle. The most effective policy initiatives for purpose ecosystem support are those that invest in enabling conditions but leave genuine space for purpose to emerge from the actors and communities involved. This means investing in trust-building processes alongside infrastructure. It means designing governance structures that are participatory and adaptive as well as accountable. It means evaluating ecosystems not only on the outputs they deliver but on the quality of the collaborative relationships and shared commitments they generate over time. And it means accepting that purpose cannot be prescribed from above, only cultivated from within.

The theoretical understanding of purpose ecosystems developed in this section connects with several adjacent literatures. We can refer to Kania and Kramer (2011) who focus on what they call collective impact, that is, cross-sector collaboration for large-scale social change. They refer to a common agenda, shared measurement systems, mutually reinforcing activities, continuous communication, and backbone support organizations as conditions for social impact. These can match with governance requirements of purpose ecosystems. Moreover, we can draw from the concept of *systemic change* in the social innovation literature (Mulgan, 2019; Westley et al., 2011). This offers a way of understanding how the activities of purpose ecosystems generate not only direct outcomes, but also deeper shifts in norms, policies, power relations, and institutional architectures. We can also build on the literature on *mission-oriented innovation policy* (Mazzucato, 2018; Schot & Steinmueller, 2018) which explain how public institutions can create enabling conditions for purpose ecosystem emergence and scale. At the same time, it remains attentive to the distinction between creating conditions and prescribing outcomes.

Purpose ecosystems differ from business ecosystems in their explicit commitment to social and environmental value creation as a primary rather than residual concern. They differ from innovation ecosystems in their emphasis on participatory and inclusive dimensions of value co-creation rather than on the alignment of complementary technological components. They differ from platform ecosystems because they reject the concept of value accumulation by a single actor and they insist on equitable value distribution. They differ from entrepreneurial ecosystems because their orientation is toward systemic social transformation rather than new venture creation. And they differ from knowledge ecosystems in their emphasis on actionable, socially impactful knowledge as the primary collective output.

These distinctions reflect fundamentally different governance logics, different accountability relationships, and different theories of how social and environmental change is generated. They also reflect a different understanding of what purpose is and how it comes into being. In purpose ecosystems, purpose is not an input that drives the system. It grows from within and shaped by the interactions among actors. The next section synthesizes these insights into a coherent framework for understanding and designing purpose ecosystems. The perspective assumed is that of an emergent conception of purpose. This choice appears more consistent with the nature of the phenomena under observation even if it entails analytical limitations in terms of immediate measurability.

Box 1.2 Purpose: Deliberated or Emergent?
A common misconception about purpose ecosystems is that purpose is established at the beginning, written into founding documents, and then implemented through governance structures. This volume takes a different view. Purpose in ecosystems is never fully deliberated in advance. It emerges gradually through the interaction of actors who share values, concerns, and a vision of change. Governance structures do not create purpose. At their best, they recognize it, protect it, and create conditions for it to deepen over time.

This has direct implications for policy-driven ecosystem initiatives. For instance, top-down mechanisms, such as Italian PNRR-funded ecosystems, play a valuable role in building enabling conditions. However, they cannot generate purpose itself. The most effective policy designs are those that invest in enabling conditions, leaving genuine space for purpose to emerge from the actors and communities involved. Policy creates the container. Purpose fills it from within.

Spotlight 1.1 The Mondragon Cooperative Corporation

The Mondragon Cooperative Corporation was not designed as an ecosystem. It emerged from a crisis. In 1941, a young Basque priest named José María Arizmendiarrieta arrived in the small town of Mondragón, in the industrial heartland of the Basque Country, to find a community devastated by the Spanish Civil War with widespread unemployment, deep poverty, and a generation of young people with no visible future. Rather than delivering charity, Arizmendiarrieta invested in education. In 1943, he founded a small polytechnic school, funded by donations from local families, where workers' children could develop the skills the territory needed. Twelve years later, five of his students left the factories where they worked to found ULGOR, the first cooperative enterprise. Arizmendiarrieta's goal was not to create a cooperative. His goal was social transformation. The cooperative was the instrument, not the end.

What makes Mondragon a purpose ecosystem is precisely this sequence, purpose first, structure second. The shared values that motivated the founders (dignity of labor, democratic participation, solidarity, and social responsibility) were not the product of a governance document or a strategic plan. They emerged from the experience of a community that had suffered together and decided to build something different together. The organizational structures that followed (the cooperative bank Caja Laboral, the social welfare body Lagun Aro, the research centers, and eventually Mondragon University) were responses to needs that the ecosystem generated as it grew, not blueprints imposed from above.

Today Mondragon comprises more than 100 federated cooperatives employing over 80,000 people across industry, finance, retail, and education. The ecosystem's governance, based on the principle of one person, one vote in each cooperative's General Assembly, produces a salary ratio of approximately 1:6 between the lowest and highest earners. When cooperatives face economic difficulties, workers accept salary reductions before jobs are cut. When one cooperative grows, it helps others to develop. This is the expression of shared values that have constituted the ecosystem's identity since 1956.

Mondragon reveals the anatomy of a purpose ecosystem. Its enabling conditions were present in the Basque territory from the beginning. We refer to high social capital, dense civic culture, and strong associational life. Its Transformative Purpose (social transformation through democratic economic participation) was not articulated in advance but emerged through practice and deepened over decades. Its governance is genuinely non-hierarchical. No single actor directs the ecosystem, and purpose is maintained through the shared identity of cooperative membership rather than through surveillance. The lesson it offers is simple, the most durable purpose ecosystems are those that begin with the most authentic shared values. Structure, technology, and resources are amplifiers. They cannot substitute for the purpose that gives them meaning.

1.3 The Purpose Ecosystem Compass

The Purpose Ecosystem Compass (PEC) is an integrative framework that provides orientation, not prescription, for purpose ecosystems. A compass does not tell you exactly where to go. It helps you orient yourself, understand your direction, and navigate in conditions of uncertainty. This is what the PEC does for purpose ecosystems. It identifies the four fundamental dimensions that every purpose ecosystem must address and shows how these dimensions relate to each other. It does not offer a step-by-step implementation guide, because purpose ecosystems are too varied, too context-dependent, and too dynamic for that kind of guidance to be useful. What the PEC offers is a shared language and a shared map.

The conceptual framework proposed does not emerge from a single, unified design at the outset. It has been developed through successive reformulations, in dialogue with empirical cases, institutional constraints, and the limitations identified in existing theoretical paradigms.

The PEC draws on several theoretical traditions: ecosystem and platform theory (Iansiti & Lakhani, 2020; Jacobides et al., 2018), complex adaptive systems (Holland, 1992; Stacey, 1995), social value creation (Emerson, 2003; Mulgan, 2010), mission-oriented innovation policy (Mazzucato, 2018), and impact frameworks aligned with the Sustainable Development Goals (Sachs et al., 2019). It is designed to serve two purposes simultaneously. As an analytical tool, it helps us understand how existing purpose ecosystems generate (or fail to generate) social, environmental, and broader societal impact. As a normative guide, it specifies the structural and process requirements that purpose ecosystem designers must attend to in order to build systems capable of sustained transformative action.

The PEC has four components, corresponding to the four cardinal points of a compass: Enabling Conditions, Digital Mechanisms, Transformative Purpose, and Governance Architecture (Fig. 1.1).

Each component captures a distinct dimension of purpose ecosystem functioning. Together, they form an integrated system in which each component shapes and depends on the others. No single component is sufficient on its own.

A purpose ecosystem generates meaningful and durable impact when all four components are present, well-developed, and aligned with each other. The failure of any one component reduces the effectiveness of all the others. Strong enabling conditions without adequate governance cannot ensure that collective resources are directed toward genuine social and environmental purposes. Sophisticated digital tools without enabling conditions will simply not reach the communities that need

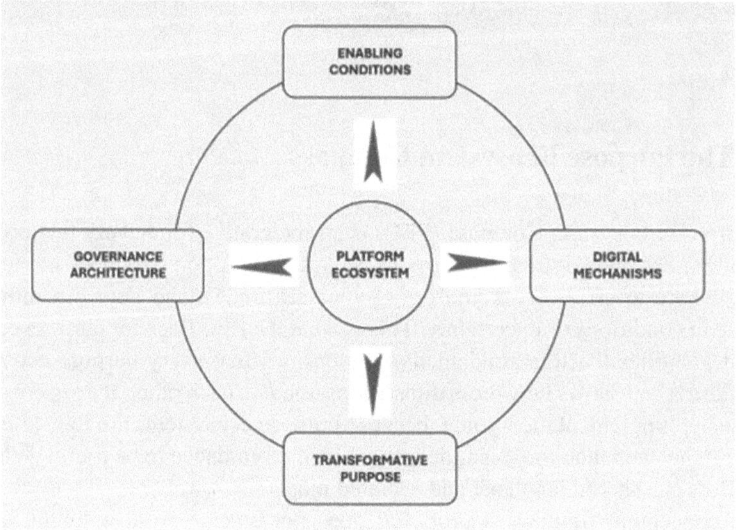

Fig. 1.1 The purpose ecosystem compass. *Source* Authors' own elaboration

them most. A genuine Transformative Purpose without accountability mechanisms cannot be verified, sustained, or protected from purpose-washing.

Enabling Conditions are the structural and relational foundation that make it possible for a purpose ecosystem to form, grow, and sustain itself over time. They include physical and digital infrastructure, institutional quality, financial resources, knowledge capacity, and the social capital that allows diverse actors to trust and collaborate with each other. Chapter 2 examines enabling conditions in depth, developing the Purpose Ecosystem Propensity Index (PEPI), a composite indicator that measures territorial readiness for purpose ecosystem development. It also examines how institutional intervention can support purpose ecosystem development in territories where enabling conditions are weak, maintaining at the same time the fundamental distinction between what policy can create and what it cannot. Policy creates the container, but purpose must fill it from within.

Therefore, the PEPI is the operational instrument through which the component "Enabling Conditions" becomes empirically measurable and territorially grounded. If the PEC tells us *what* a purpose ecosystem needs in order to function, the PEPI tells us *how much* of that foundation a given territory already has in place.

Some dimensions of enabling conditions deserve particular attention. The foundational infrastructure is relevant. We refer to the physical connectivity, technical interoperability, and human capabilities that allow actors to participate meaningfully in ecosystem activities. Without these foundations, participation is structurally impossible for large parts of the population that purpose ecosystems most need to reach (Warschauer, 2003). Moreover, institutional readiness is an important enabling condition. We refer to the regulatory frameworks, financing mechanisms, and organizational capabilities that allow ecosystems to operate effectively within their broader institutional context. Among enabling conditions, also relational trust is significant. This is the accumulated confidence among ecosystem participants that others will behave with honesty, competence, and genuine commitment to shared purpose. Trust cannot be designed into an ecosystem from the outside. It develops through patterns of behavior over time, through transparency, through equitable sharing of benefits and risks, through genuine community engagement, and through demonstrated accountability to shared commitments (Coleman, 1990; Ostrom, 2010).

Transformative Purpose is the component that answers the most fundamental question about any purpose ecosystem: why does it exist? It captures the social and environmental direction toward which all ecosystem activities are oriented. It is not a measured result. It is a compass direction, the shared commitment that motivates ecosystem actors, guides collective decisions, and gives meaning to everything the ecosystem does. Without a genuine Transformative Purpose, an ecosystem may be well-structured, well-governed, and digitally sophisticated, but it is not a purpose ecosystem. It is simply a network. Transformative Purpose in the PEC is related to the commitment to generating social value for communities, reducing inequalities, expanding opportunities, supporting civic participation, and building the conditions for human flourishing in all its dimensions. It encompasses empowerment, community agency, cultural vitality, well-being, and the structural conditions that determine what is possible for people in a given territory. Moreover, purpose can be related to environmental sustainability, that is, the commitment to managing natural resources responsibly, reducing greenhouse gas emissions, protecting biodiversity, supporting the circular economy, and contributing to the ecological transitions that communities and future generations depend on. Such dimensions are equally important and mutually reinforcing. A purpose ecosystem that pursues social transformation and degrades the natural environment is contradicting itself. One that protects the environment and generates social inequalities is equally incomplete.

Digital Mechanisms play important roles within purpose ecosystems. They reduce coordination costs and help diverse actors find and connect with each other by matching platforms, resource-sharing systems, and collaborative working environments all reduce the frictions that would otherwise prevent purpose ecosystems from assembling the complementary capabilities they need. They create informational transparency, making visible the environmental and social consequences of ecosystem activities and enabling forms of accountability that were previously impossible to achieve. And they support learning and adaptation by data analytics, feedback systems, and adaptive management tools enable ecosystems to continuously improve their strategies in response to evidence about what is working and what is not. Chapter 4 examines digital mechanisms in detail, analyzing their three core functions, the two different pathways through which they develop in emergent versus policy-initiated ecosystems, and the fundamental principle of amplification without substitution. Digital tools amplify what already exists, but they cannot generate the shared values and relational trust on which authentic purpose ecosystems depend. We need to underline that digital tools are means, not ends. Their value depends entirely on what purposes they are made to serve and how well they are governed. The same technology that enables community empowerment in one context can produce surveillance and exclusion in another. Additionally, the deployment of digital tools always requires adequate enabling conditions to be present first. Deploying sophisticated platforms in communities without adequate connectivity, digital literacy, or institutional trust will not generate leverage. It will generate frustration, exclusion, and a further erosion of the trust that purpose ecosystem development depends on.

Governance Architecture is the component that holds the other three together. It refers to the rules, processes, accountability mechanisms, and institutional arrangements through which the ecosystem makes collective decisions, maintains alignment with its shared purpose, and remains accountable to the communities it serves. Chapter 5 develops this component in full detail, examining how governance works differently in organically emergent ecosystems and in policy-initiated ecosystems, and addressing the deeper theoretical questions that all forms of purpose ecosystem governance must confront.

Governance in purpose ecosystems faces a distinctive challenge that sets it apart from governance in conventional firms or public institutions. Purpose ecosystems are not hierarchically controlled, which means no single actor has the authority to direct the activities of all others. They are not purely market coordinated either, which means the social and environmental missions that define their identity cannot be adequately governed by price signals and contractual enforcement alone. They require a third mode of governance based on shared values, participatory processes, mutual accountability, and the co-construction of purpose described in Sect. 1.2. This mode of governance is more demanding than either hierarchical or market coordination, but it is also more robust, because it generates the genuine commitment and collective identity that sustained transformative action requires.

Four governance functions are essential in purpose ecosystems. Purpose recognition describes the processes through which the ecosystem's emerging shared mission is made visible, given voice, and kept alive over time. Unlike purpose articulation in conventional organizations (where purpose is defined by leadership and communicated downward), in purpose ecosystems this function works in the opposite direction, creating participatory spaces, deliberative forums, and community dialogues through which a purpose, that is, already latent among actors becomes recognized, shared, and progressively deepened. Participation design describes the institutional arrangements through which diverse stakeholders are enabled to contribute to and shape ecosystem activities. Accountability enforcement describes the monitoring, verification, and sanctioning mechanisms that ensure ecosystem actors behave in accordance with their shared commitments. And adaptive governance describes the learning and revision processes through which the ecosystem updates its strategies, rules, and architecture in response to evidence, changing contexts, and evolving needs. All four functions are examined in Chap. 5.

The PEC has important implications for how we understand ecosystem failure as well as success. When enabling conditions are weak, participation is structurally limited and the ecosystem cannot reach those who most need its services. When digital mechanisms are poorly designed or misaligned with ecosystem purpose, technology reproduces existing inequalities rather than challenging them. When Transformative Purpose is not systematically verified and the ecosystem loses sight of the communities it serves, accountability and direction erode. And when governance architecture is weak or captured by powerful interests, purpose gradually erodes into rhetoric. Understanding these failure modes is as analytically important as understanding the conditions for success.

The structure of the PEC also reflects an important principle about the non-linear nature of purpose ecosystem development. There is no sequence of steps that guarantees success. Enabling conditions, digital mechanisms, Transformative Purpose, and governance all develop simultaneously and in constant interaction with each other. Trust develops through governance experience. Governance improves through learning about outcomes. Digital mechanisms become more effective as enabling conditions strengthen. And outcomes improve as all four components become more aligned. This circular, iterative, and deeply relational character of purpose ecosystem development is what makes the compass metaphor apt. Navigation is not a one-time act. It is a continuous practice of orientation, adjustment, and learning.

The purpose ecosystem offers a new grammar for organizing collaboration, orchestrating innovation, and governing transitions in conditions of deep uncertainty and interdependence. It is not a blueprint. It is not a technology. It is an organizational form that, when well-designed and well-governed, can generate social, environmental, and broader societal impact that no single actor could achieve alone. The PEC provides the analytical architecture for understanding how this form works, what it needs to thrive, and how we can evaluate whether it is genuinely fulfilling its purpose.

Spotlight 1.2 The RAISE Ecosystem in Liguria

When RAISE (Robotics and AI for Socio-economic Empowerment) was established in Liguria in 2022, funded by Italy's National Recovery and Resilience Plan with nearly 110 million euros, it was formally institutionalizing a Transformative Purpose that the Ligurian territory had been developing organically for decades. It was the application of advanced technology to genuinely improve the quality of human life. The Italian Institute of Technology, the University of Genoa, and the National Research Council had built world-class research capabilities in robotics and artificial intelligence, generating what observers call a "Robot Valley" centered on Genoa. But what makes RAISE more than a research program is the direction in which that capability is explicitly oriented.

Liguria is a region that has confronted the challenges of demographic aging, industrial decline, and geographic complexity earlier and more acutely than almost any other European territory. These challenges, rather than being treated as obstacles to development, became the generative conditions for a distinctive technological purpose that was developing robotics and AI that serve people's real needs, particularly the needs of an aging population, workers in complex industrial environments, and a port system that needs to become more sustainable. The six application areas around which RAISE organizes its 26 partner organizations (healthcare and human care, inclusive urban environments, environmental monitoring and risk prevention, smart and sustainable ports, technology transfer, and education) reflect a territorial reading of where technology can genuinely improve people's lives.

The Spoke dedicated to healthcare develops robotic prosthetics, exoskeletons, and assistive devices for independent living. The Spoke on inclusive urban environments designs solutions to make cities genuinely navigable for people with physical and sensory limitations. The Spoke on environmental monitoring builds systems to anticipate natural risks in a territory highly vulnerable to floods and landslides. Its training program, which delivered 973 h of activities reaching over 7600 participants including students, researchers, and professionals, demonstrates a genuine commitment to developing the territorial capabilities that sustain purpose beyond the Italian PNRR funding period. Female participation reached 52% of the total, exceeding targets and confirming the ecosystem's commitment to gender balance in technological disciplines.

Reading RAISE through the framework of this chapter reveals the strengths and the challenges of policy-initiated purpose ecosystems. Its formal governance structure provides the institutional scaffolding that enables coordination and accountability among 26 diverse partners. Its cascading calls have expanded the ecosystem's reach to include more than 40 affiliated enterprises. The central question RAISE poses is whether the Transformative Purpose of improving people's lives through technology is genuinely constitutive of the ecosystem's identity, or whether it is a legitimating narrative overlaid on activities primarily driven by research excellence. What is clear is that RAISE has created the structural conditions within which authentic purpose can develop. Such conditions are multi-actor collaboration, community engagement, territorial anchoring, and a directional commitment that goes beyond disciplinary boundaries.

References

Adner, R. (2006). Match your innovation strategy to your innovation ecosystem. *Harvard Business Review, 84*(4), 98–107.

Adner, R. (2017). Ecosystem as structure: An actionable construct for strategy. *Journal of Management, 43*(1), 39–58.

Adner, R., & Kapoor, R. (2010). Value creation in innovation ecosystems: How the structure of technological interdependence affects firm performance in new technology generations. *Strategic Management Journal, 31*(3), 306–333.

Alvedalen, J., & Boschma, R. (2017). A critical review of entrepreneurial ecosystems research: Towards a future research agenda. *European Planning Studies, 25*(6), 887–903.

Arthur Tansley, A. G. (1935). The use and abuse of vegetational concepts and terms. *Ecology, 16*(3), 284–307.

Audretsch, D. B., Cunningham, J. A., Kuratko, D. F., Lehmann, E. E., & Menter, M. (2019). Entrepreneurial ecosystems: Economic, technological, and societal impacts. *The Journal of Technology Transfer, 44*, 313–325

Autio, E., & Thomas, L. D. W. (2020). Innovation ecosystems in management: An organizing typology, In *Oxford Encyclopedia of Business and Management*. Oxford University Press.

Autio, E., Nambisan, S., Thomas, L. D. W., & Wright, M. (2018). Digital affordances, spatial affordances, and the genesis of entrepreneurial ecosystems. *Strategic Entrepreneurship Journal, 12*(1), 72–95.

Axelrod, R., & Cohen, M. D. (1999). Harnessing Complexity: Organizational Implications of a Scientific Frontier. Free Press.

Bahrami, H., & Evans, S. (1995). Flexible re-cycling and high-technology entrepreneurship. *California Management Review, 37*(3), 62–89.

Baldwin, C. Y., & Woodard, C. J. (2009). The architecture of platforms: A unified view. In A. Gawer (Ed.), *Platforms, Markets and Innovation* (pp. 19–44). Edward Elgar.

Barney, J. (1991). Firm resources and sustained competitive advantage. *Journal of Management, 17*(1), 99–120.

Bartlett, C. A., & Ghoshal, S. (1994). Changing the role of top management: Beyond strategy to purpose. *Harvard Business Review, 72*(6), 79–88.

Baumol, W. J. (1990). Entrepreneurship: Productive, unproductive, and destructive. *Journal of Political Economy, 98*(5), 893–921.

Baumol, W. J. (1993). Entrepreneurship, Management, and the Structure of Payoffs. MIT Press.

Bejjani, M., Göcke, L., & Menter, M. (2023). Digital entrepreneurial ecosystems: A systematic literature review. *Technological Forecasting and Social Change, 189*, 122372.

Bourdieu, P. (1986). The forms of capital. In J. G. Richardson (Ed.), *Handbook of Theory and Research for the Sociology of Education* (pp. 241–258). Greenwood.

Cantner, U., Cunningham, J. A., Lehmann, E. E., & Menter, M. (2021). Entrepreneurial ecosystems: A dynamic lifecycle model. *Small Business Economics, 57*, 407–423.

Carayannis, E. G., & Campbell, D. F. J. (2009). 'Mode 3' and 'Quadruple Helix': Toward a 21st century fractal innovation ecosystem. *International Journal of Technology Management, 46*(3/4), 201–234.

Cennamo, C., & Santaló, J. (2013). Platform competition: Strategic trade-offs in platform markets. *Strategic Management Journal, 34*(11), 1331–1350.

Cennamo, C. (2019). Competing in digital markets: A platform-based perspective. *Academy of Management Perspectives, 35*(2), 265–291.

Chesbrough, H., Kim, S., & Agogino, A. (2014). Chez Panisse: Building an open innovation ecosystem. *California Management Review, 56*(4), 144–171.

Clarysse, B., Wright, M., Bruneel, J., & Mahajan, A. (2014). Creating value in ecosystems: Crossing the chasm between knowledge and business ecosystems. *Research Policy, 43*(7), 1164–1176.

Cohen, B. (2006). Sustainable valley entrepreneurial ecosystems. *Business Strategy and the Environment, 15*(1), 1–14.

Coleman, J. S. (1990). *Foundations of social theory*. Harvard University Press.
Corrente, S., Greco, S., Nicotra, M., Romano, M., & Schillaci, C. (2019). Evaluating and comparing entrepreneurial ecosystems using SMAA and SMAA-S. *The Journal of Technology Transfer, 44*(2), 485–519.
Crane, A., Palazzo, G., Spence, L. J., & Matten, D. (2014). Contesting the value of "creating shared value". *California Management Review, 56*(2), 130–153.
Cunningham, J. A., Menter, M., & Wirsching, K. (2019). Entrepreneurial ecosystem governance: A principal investigator-centered governance framework. *Small Business Economics, 52*, 545–562.
Dattée, B., Alexy, O., & Autio, E. (2018). Maneuvering in poor visibility: How firms play the ecosystem game when uncertainty is high. *Academy of Management Journal, 61*(2), 466–498.
DiMaggio, P. J., & Powell, W. W. (1983). The iron cage revisited: Institutional isomorphism and collective rationality in organizational fields. *American Sociological Review, 48*(2), 147–160.
Elkington, J. (1997). Cannibals with Forks: The Triple Bottom Line of 21st Century Business. Capstone.
Emerson, K., & Nabatchi, T. (2015). *Collaborative Governance Regimes*. Georgetown University Press.
Evans, D. S., & Schmalensee, R. (2016). Matchmakers: The New Economics of Multisided Platforms. *Harvard Business Review Press*.
Feld, B. (2012). *Startup Communities: Building an Entrepreneurial Ecosystem in Your City*. Wiley.
Florida, R. (1995). Toward the learning region. *Futures, 27*(5), 527–536.
Florida, R. (2002). *The Rise of the Creative Class and How It's Transforming Work, Leisure, Community and Everyday Life*. New York: Basic Books.
Freeman, R. E. (1984). *Strategic management: A stakeholder approach*. Pitman.
Gartenberg, C., & Yiu, S. (2023). Acquisitions and corporate purpose. *Strategy Science, 8*(4), 444–463.
Gawer, A., & Cusumano, M. A. (2002). *Platform Leadership: How Intel, Microsoft, and Cisco Drive Industry Innovation*. Harvard Business School Press.
Gawer, A., & Cusumano, M. A. (2008). How companies become platform leaders. *MIT Sloan Management Review, 49*(2), 28–35.
Gawer, A., & Cusumano, M. A. (2014). Industry platforms and ecosystem innovation. *Journal of Product Innovation Management, 31*(3), 417–433.
Gawer, A., & Phillips, N. (2013). Institutional work as logics shift: The case of Intel's transformation to platform leader. *Organization Studies, 34*(8), 1035–1071.
George, G., Haas, M. R., McGahan, A. M., Schillebeeckx, S. J. D., & Tracey, P. (2023). Purpose in the for-profit firm: A review and framework for management research. *Journal of Management, 49*(6), 1841–1869.
Hannah, D. P., & Eisenhardt, K. M. (2018). How firms navigate cooperation and competition in nascent ecosystems. *Strategic Management Journal, 39*(12), 3163–3192.
Holland, J. H. (1992). *Adaptation in Natural and Artificial Systems. MIT Press*.
Hollensbe, E., Wookey, C., Hickey, L., George, G., Nichols, V., & George, G. (2014). Organizations with purpose: From the editors. *Academy of Management Journal, 57*(6), 1227–1234.
Holling, C. S. (1973). Resilience and stability of ecological systems. *Annual Review of Ecology and Systematics, 4*, 1–23.
Hurth, V., Ebert, C., & Prabhu, J. (2018). *Organisational purpose: The construct and its antecedents and consequences*. Cambridge Judge Business School Working Paper, No. 02/2018.
Iansiti, M., & Levien, R. (2004). Strategy as ecology. *Harvard Business Review, 82*(3), 68–78.
Iansiti, M., & Lakhani, K. R. (2020). Competing in the Age of AI: Strategy and Leadership When Algorithms and Networks Run the World. *Harvard Business Review Press*.
Isenberg, D. J. (2010). How to start an entrepreneurial revolution. *Harvard Business Review, 88*(6), 40–50.
Isenberg, D. J. (2011). The Entrepreneurship Ecosystem Strategy as a New Paradigm for Economic Policy. *Babson Entrepreneurship Ecosystem Project*.
Jacobides, M. G., Cennamo, C., & Gawer, A. (2018). Towards a theory of ecosystems. *Strategic Management Journal, 39*(8), 2255–2276.
Järvi, K., Almpanopoulou, A., & Ritala, P. (2018). Organization of knowledge ecosystems: Prefigurative and partial forms. *Research Policy, 47*(8), 1523–1537.

Kania, J., & Kramer, M. (2011). Collective impact. *Stanford Social Innovation Review, 9*(1), 36–41.
Levie, J., & Autio, E. (2008). A theoretical grounding and test of the GEM model. *Small Business Economics, 31*(3), 235–263.
Lewin, A. Y., & Volberda, H. W. (1999). Prolegomena on coevolution: A framework for research on strategy and new organizational forms. *Organization Science, 10*(5), 519–534.
Mackey, J., & Sisodia, R. (2013). *Conscious Capitalism: Liberating the Heroic Spirit of Business.* Harvard Business Review Press.
Mazzucato, M. (2018). *The value of everything: Making and taking in the global economy.* Allen Lane.
Moore, J. F. (1993). Predators and prey: A new ecology of competition. *Harvard Business Review, 71*(3), 75–86.
Moore, J. F. (1996). The Death of Competition: Leadership and Strategy in the Age of Business Ecosystems. *Harper Business.*
Morrison, A. D., & Mota, R. (2021). A theory of organizational purpose. *Academy of Management Review, 48*(2), 203–219
Mulgan, G. (2010). Measuring social value. *Stanford Social Innovation Review, 8*(3), 38–43.
Mulgan, G. (2019). *Social innovation: How societies find the power to change.* Policy Press.
Nambisan, S., Wright, M., & Feldman, M. (2019). The digital transformation of innovation and entrepreneurship: Progress, challenges and key themes. *Research Policy, 48*(8), Article 103773.
Neck, H. M., Meyer, G. D., Cohen, B., & Corbett, A. C. (2004). An entrepreneurial system view of new venture creation. *Journal of Small Business Management, 42*(2), 190–208.
Nicotra, M. (2019). *Shifting paradigms in entrepreneurial ecosystems: A data-driven perspective.* UTET.
Nicotra, M., & Romano, M. (2018). *Ecosistemi della Conoscenza e Absorptive Capacity.* Giappichelli.
Nicotra, M., Romano, M., Del Giudice, M., & Schillaci, C. E. (2018). The causal relation between entrepreneurial ecosystem and productive entrepreneurship. *The Journal of Technology Transfer, 43*(3), 640–673.
North, D. C. (1990). *Institutions, institutional change and economic performance.* Cambridge University Press.
Ostrom, E. (2010). Beyond markets and states: Polycentric governance of complex economic systems. *American Economic Review, 100*(3), 641–672.
Pickett, S. T. A., & Cadenasso, M. L. (2002). The ecosystem as a multidimensional concept: Meaning, model, and metaphor. *Ecosystems, 5*(1), 1–10.
Porter, M. E. (1990). *The competitive advantage of nations.* Free Press
Porter, M. E. (1998). Clusters and the new economics of competition. *Harvard Business Review, 76*(6), 77–90.
Provan, K. G., & Kenis, P. (2008). Modes of network governance: Structure, management, and effectiveness. *Journal of Public Administration Research and Theory, 18*(2), 229–252.
Porter, M. E., & Kramer, M. R. (2011). Creating shared value. *Harvard Business Review, 89*(1/2), 62–77.
Putnam, R. D. (1993). *Making democracy work: Civic traditions in modern Italy.* Princeton University Press.
Radziwon, A., Bogers, M., & Bilberg, A. (2017). Creating and capturing value in a regional innovation ecosystem: A study of how manufacturing SMEs develop collaborative solutions. *International Journal of Technology Management, 75*(1–4), 73–96.
Rochet, J.-C., & Tirole, J. (2003). Platform competition in two-sided markets. *Journal of the European Economic Association, 1*(4), 990–1029.
Romano, M., Del Giudice, M., & Nicotra, M. (2014a). Knowledge creation and exploitation in Italian Universities: the role of internal policies for patent activity. *Journal Of Knowledge Management, 18*, 952–970.
Romano, M., Catalfo, P., & Nicotra, M. (2014b). Science parks and intellectual capital: An integrated model for intangibles' representation, evaluation and control. *Journal Of Intellectual Capital, 15*, pp. 537–553.

Romano, M., Nicotra, M., Corrente, S., & Castrogiovanni, A. (2020). Start-ups, innovation and performance: An empirical analysis using SMAA-S. In A. Caputo & M. Pellegrini (Eds.), *The entrepreneurial behaviour* (pp. 167–186). Emerald Publishing.

Sachs, J. D., Schmidt-Traub, G., Mazzucato, M., Messner, D., Nakicenovic, N., & Rockstrom, J. (2019). Six transformations to achieve the sustainable development goals. *Nature Sustainability, 2*(9), 805–814.

Schot, J., & Steinmueller, W. E. (2018). Three frames for innovation policy: R&D, systems of innovation and transformative change. *Research Policy, 47*(9), 1554–1567.

Scott, W. R. (1995). *Institutions and Organizations*. Sage.

Shipilov, A., & Gawer, A. (2020). Integrating research on interorganizational networks and ecosystems. *Academy of Management Annals, 14*(1), 92–121.

Spigel, B. (2017). The relational organization of entrepreneurial ecosystems. *Entrepreneurship Theory and Practice, 41*(1), 49–72.

Stacey, R. D. (1995). The science of complexity: An alternative perspective for strategic change processes. *Strategic Management Journal, 16*(6), 477–495.

Stam, E. (2015). Entrepreneurial ecosystems and regional policy: A sympathetic critique. *European Planning Studies, 23*(9), 1759–1769.

Teece, D. J. (2007). Explicating dynamic capabilities: The nature and microfoundations of (sustainable) enterprise performance. *Strategic Management Journal, 28*(13), 1319–1350.

Teece, D. J., Pisano, G., & Shuen, A. (1997). Dynamic capabilities and strategic management. *Strategic Management Journal, 18*(7), 509–533.

Tiwana, A., Konsynski, B., & Bush, A. A. (2010). Research commentary — Platform evolution: Coevolution of platform architecture, governance, and environmental dynamics. *Information Systems Research, 21*(4), 675–687.

Valkokari, K. (2015). Business, innovation, and knowledge ecosystems: How they differ and how to survive and thrive within them. *Technology Innovation Management Review, 5*(8), 17–24.

Wareham, J., Fox, P. B., & Cano Giner, J. L. (2014). Technology ecosystem governance. *Organization Science, 25*(4), 1195–1215.

Westley, F., Olsson, P., Folke, C., Homer-Dixon, T., Vredenburg, H., Loorbach, D., Thompson, J., Nilsson, M., Lambin, E., Sendzimir, J., Banerjee, B., Galaz, V., & van der Leeuw, S. (2011). Tipping toward sustainability: Emerging pathways of transformation. *Ambio, 40*(7), 762–780.

Willis, A. J. (1997). The ecosystem: An evolving concept viewed historically. *Functional Ecology, 11*(2), 268–271.

Zucker, L. G., Darby, M. R., & Brewer, M. B. (1998). Intellectual human capital and the birth of U.S. biotechnology enterprises. *American Economic Review, 88*(1), 290–306.

Chapter 2
Enabling Conditions

Purpose ecosystems do not emerge from blueprints. They are not assembled by design committees, activated by funding decisions, or produced by governance structures alone. They emerge from the bottom-up, from the interaction of actors who share values, concerns, and a vision of social and environmental change, in territories that provide the conditions making such interaction possible, productive, and sustained over time.

Enabling Conditions are the *structural and relational foundations* that make it possible for a purpose ecosystem to form, grow, and sustain itself. These conditions are not imposed from outside. They are properties of territories, accumulated over time through the development of institutions, knowledge systems, civic cultures, and social relationships. Some territories have them in abundance. Others have significant gaps. The distribution of enabling conditions reflects deep historical, institutional, and cultural processes that shape what is possible for collective action in each place.

This chapter develops two complementary arguments about enabling conditions. In Sect. 2.1, enabling conditions are best understood as the latent properties of territories that make purpose ecosystem emergence more or less likely, irrespective of any deliberate policy intervention. When a territory has strong institutional quality, rich knowledge infrastructure, dense social capital, and vibrant civic life, purpose ecosystems are more likely to emerge spontaneously from the interaction of actors who share a vision of change. The Purpose Ecosystem Propensity Index (PEPI), introduced in this section, provides a composite measure of this territorial propensity. In Sect. 2.2, institutional intervention is considered to play a legitimate role in supporting purpose ecosystem development, particularly in territories where enabling conditions are weak or uneven. But this role has clear limits because policy can create structural preconditions, but it cannot generate the shared values and relational trust on which authentic purpose ecosystems depend. Section 2.3 operationalizes the PEPI empirically, applying the Benefit-of-the-Doubt method to compute region-specific composite scores for all 20 Italian regions and interpreting the results in light of the theoretical typology developed in Sects. 2.1 and 2.2.

M. Nicotra and M. Romano, *Purpose Ecosystems*, Contributions to Management Science, https://doi.org/10.1007/978-3-032-38382-2_2

2.1 Enabling Conditions for Purpose Ecosystem Emergence

Why do some territories generate flourishing networks of purpose collaboration and others remain fragmented and institutionally thin? This is one of the most important and least resolved questions in the study of collective action, regional development, and social innovation. It cannot be answered by looking only at individual organizations, however capable or committed. It requires attending to the territorial context in which those organizations are embedded, namely the institutions, knowledge systems, social networks, and cultural norms that constitute the enabling environment for collective action, and particularly for the demanding, multi-stakeholder, long-term collaboration that purpose ecosystems require.

Management scholarship has developed a rich set of frameworks for understanding territorial enabling conditions, drawing on regional economics, institutional theory, social capital theory, innovation systems research, and entrepreneurial ecosystem theory. This section synthesizes these frameworks into a coherent account of what enabling conditions are, why they matter for purpose ecosystem emergence, and how they can be systematically assessed. It then introduces the Purpose Ecosystem Propensity Index (PEPI) as a practical instrument for measuring territorial enabling conditions in ways that are theoretically grounded and empirically applicable.

The insight that economic and social activity is territorially embedded is one of the most important contributions of economic geography to management scholarship. Firms, organizations, and individuals operate not in abstract markets or institutional vacuums but in specific places with specific histories, relationships, and institutional configurations. Polanyi's (1944) foundational concept of embeddedness established the principle that economic action is always embedded in social relations and cannot be understood apart from them. Granovetter's (1985) influential reformulation extended this insight to show how the performance of economic actors depends on the quality of the social networks in which they are embedded, not just on their individual capabilities or the incentives they face.

The territorial dimension of embeddedness was developed most systematically by economic geographers working in the tradition of Marshall (1890), who identified the agglomeration effects that make industrial districts more productive than equivalent concentrations of isolated firms. Marshall argued that proximity generates positive externalities through the sharing of knowledge, labor, and specialized infrastructure. These concepts have been extended through decades of research on industrial districts (Becattini, 1990; Pyke et al., 1992), clusters (Porter, 1990, 1998), regional innovation systems (Asheim & Gertler, 2005; Cooke, 1992), and learning regions (Florida, 1995; Lundvall & Johnson, 1994). What unites these diverse traditions is the recognition that place matters. The territorial context in which firms and organizations operate is not merely a background condition but an active determinant of their capacity for innovation, collaboration, and value creation.

Purpose ecosystems are not only embedded in economic and institutional territories. They are embedded in communities, with their specific histories of collaboration and conflict, their cultural identities and values, their patterns of trust and distrust, and their accumulated stocks of social capital. The territorial conditions that matter for purpose ecosystem emergence therefore include not only the institutional and economic infrastructures, but also the relational and cultural fabric. Understanding territorial enabling conditions for purpose ecosystems requires integrating both dimensions.

Institutions are the most fundamental enabling condition for collective action of any kind. North (1990) defines them as "the rules of the game in a society." They mainly refer to laws, regulations, property rights, and contract enforcement mechanisms. The quality of these institutional arrangements, their credibility, consistency, impartiality, and resistance to capture by powerful interests determines the basic conditions under which collective action is possible.

The institutional economics literature, developed by Acemoglu et al. (2001), North (1990), Rodrik et al. (2004), and others, has consistently demonstrated that institutional quality is among the strongest predictors of long-term economic and social performance across territories and nations. High-quality institutions reduce the transaction costs of economic exchange, protect against arbitrary expropriation by powerful actors, create predictable regulatory environments that support long-term investment, and build the public trust that enables voluntary compliance with collective rules. For purpose ecosystems, which depend on the voluntary commitment of diverse actors to shared goals over extended periods, institutional quality is especially critical. Ecosystems that emerge in territories with weak, inconsistent, or captured institutions face elevated risks of governance failure, resource misallocation, and purpose erosion.

Institutional complementarities, developed by Hall and Soskice (2001) in the comparative capitalisms literature and extended to the innovation systems context by Amable (2003) and Whitley (1999), add an important dimension to the analysis. Institutions do not operate in isolation. Their effectiveness depends on how well they fit together with other institutions in the same territory. A territory with strong formal property rights but weak social trust may find that formal enforcement is too costly to enable the relational governance that purpose ecosystems require. A territory with strong social norms of reciprocity but weak formal contract enforcement may find that informal governance is insufficient to sustain the scale and complexity of multi-actor purpose ecosystem activities. Effective institutional enabling conditions require not only individual institutional quality but institutional complementarity. The coherence and mutual reinforcement of formal and informal institutions can reduce the costs of collective action.

The European Quality of Government Index (EQI), developed by Charron et al. (2014, 2019) at the University of Gothenburg, provides the most comprehensive and regionally comparable measure of institutional quality currently available for European territories. The EQI captures four dimensions of government quality that are quality of public services, quality of public administration, corruption control, and rule of law. It is assessed through a combination of citizen surveys and expert assessments at the NUTS2 regional level. This regional granularity is essential for purpose ecosystem analysis. Institutional quality varies enormously within countries as well as between them, and it is the quality of regional and local institutions (rather than national averages) that most directly shapes the enabling environment for territorial purpose ecosystems.

In addition, knowledge is the most fundamental input to the purpose problem-solving that defines purpose ecosystem activity. Purpose ecosystems address complex social and environmental challenges. They require sophisticated analytical capabilities, evidence-based decision-making, continuous learning from experience, and the capacity to generate and diffuse innovations that can change the conditions producing the problems being addressed. The knowledge infrastructure of a territory (including its universities, research centers, technology transfer institutions, and knowledge-intensive enterprises) determines whether these capabilities are available within the ecosystem's geographic reach.

Regional innovation systems research, pioneered by Cooke (1992) and developed by Asheim and Gertler (2005), Lundvall (1992), and Nelson (1993), has established that innovation is fundamentally a systemic and territorial phenomenon. Firms and organizations do not innovate in isolation. Their innovation capacity depends on their ability to access and absorb knowledge generated by universities, public research institutions, and other firms, to collaborate with partners who bring complementary knowledge and capabilities, and to participate in the knowledge flows that circulate through territorial networks of interaction. The quality and density of these knowledge flows, and the institutional frameworks that govern them, determine the innovation capacity of a territory as a whole. This in turn determines the capacity of purpose ecosystems embedded in that territory to generate the novel solutions that social and environmental transformation requires.

Cohen and Levinthal's (1990) concept of absorptive capacity, originally developed at the firm level, has been extended to the territorial level by Boschma (2005) and others. It describes the capacity of a territory to recognize, assimilate, and apply external knowledge to its own development challenges. Territorial absorptive capacity depends not only on the stock of knowledge assets (measured by R&D expenditure, number of researchers, or publication output) but on the quality of the linkages and knowledge transfer mechanisms that connect knowledge producers with knowledge users. The dynamics of territorial absorptive capacity, including the co-evolutionary adaptation of actors around shared knowledge production and the role of science parks and intermediary organizations, have been examined in depth in the Italian context (Nicotra & Romano, 2018; Romano et al., 2014; Schillaci et al., 2013). Purpose ecosystems benefit from high territorial absorptive capacity because they depend on the ability to apply the best available knowledge from multiple disciplines to complex, multi-dimensional social and environmental challenges. A territory that generates significant research output but lacks effective mechanisms for connecting researchers with practitioners, civic organizations, and community members will have lower effective absorptive capacity than its raw knowledge assets would suggest.

Related variety, developed by Boschma and Iammarino (2009) and Frenken et al. (2007), provides a further dimension of knowledge infrastructure analysis, that is, particularly relevant for purpose ecosystems. Related variety refers to the presence, in a territory, of knowledge domains that are sufficiently similar to enable cross-fertilization and recombination but sufficiently different to generate novelty through their interaction. Purpose ecosystems address challenges that span multiple domains (social services, environmental management, economic development, governance, and cultural vitality) and benefit particularly from territories where diverse but interconnected knowledge communities coexist. The presence of related variety increases the probability that purpose ecosystem actors will be able to draw on the full range of knowledge needed for multi-dimensional social and environmental innovation.

Of all the enabling conditions for purpose ecosystem emergence, social capital is the most important and difficult to build through deliberate intervention. Social capital, as defined by Coleman (1990), refers to networks, norms, and trust that facilitate action and cooperation for mutual benefit. Putnam's (1993, 2000) empirical research demonstrated that differences in social capital across Italian regions, accumulated over centuries of different historical experience, were among the strongest predictors of institutional performance, economic development, and civic vitality in contemporary Italy. Territories with high social capital, characterized by dense networks of civic associations, strong norms of reciprocity, and generalized trust among citizens, consistently outperformed territories with low social capital on virtually every dimension of collective performance that Putnam examined.

For purpose ecosystems, social capital is not merely a predictor of performance; it is a constitutive element of what purpose ecosystems are. The co-creation logic that is central to purpose ecosystem identity, as developed in Chap. 1, requires trust among diverse actors who do not share organizational affiliation, common interests, or prior history of collaboration. Without trust, the sharing of knowledge, resources, and vulnerabilities that multi-actor collaboration requires is simply too costly. Without norms of reciprocity, the sustained commitment to shared purpose that distinguishes purpose ecosystems from opportunistic partnerships cannot be maintained over time. Without civic networks, the diverse actors needed for purpose ecosystem formation cannot find each other, build relationships, and discover their shared values.

Putnam's distinction between bonding social capital, which connects similar individuals within tight-knit groups, and bridging social capital, which connects diverse individuals and groups across social boundaries, is particularly important for purpose ecosystem analysis. Bonding social capital provides the strong ties and deep trust that sustain long-term commitment among core ecosystem actors. But it is bridging social capital, the weaker but broader connections that span sectoral, cultural, and institutional boundaries, that enables the formation of the diverse, multi-stakeholder configurations that purpose ecosystems require. Territories with high bridging social capital are better positioned to generate purpose ecosystems because they have the relational component for connecting actors who would not otherwise encounter each other.

Fukuyama's (1995) analysis of trust as a form of social capital extends the concept beyond the civic associations that Putnam emphasizes to encompass the broader cultural norms that shape how people relate to strangers, institutions, and collective enterprises. High-trust societies have a structural advantage in generating large-scale collective action because their members are willing to cooperate with actors they do not personally know, to invest in shared institutions, and to honor commitments even when not directly monitored. Low-trust societies face persistent collective action problems that require expensive formal enforcement mechanisms to overcome. For purpose ecosystems operating at the territorial scale, where actors must coordinate across organizational, sectoral, and cultural boundaries over extended periods, the level of generalized trust in the territory is a fundamental enabling condition that cannot be easily replicated through formal governance mechanisms.

Woolcock's (1998) distinction between bonding, bridging, and linking social capital adds a third dimension that is important for purpose ecosystem analysis. Linking social capital refers to connections between individuals and groups with different levels of power and social status, particularly connections between community actors and institutional actors. For purpose ecosystems that aspire to genuinely co-create social value with marginalized communities rather than merely delivering services to them, linking social capital is essential. It determines whether community members have genuine access to the institutional resources and decision-making processes of the ecosystem, or whether they remain peripheral to an ecosystem that claims to serve them. Territories with high linking social capital, characterized by accessible public institutions, responsive civic infrastructure, and cultural norms, are better positioned to generate purpose ecosystems that are genuinely inclusive and community-anchored.

The associational life of a territory, the density and diversity of voluntary organizations, civic groups, cultural associations, and social movements through which citizens engage in collective action, is a manifestation and a generator of social capital. Dense associational life provides the organizational infrastructure through which social capital is produced and reproduced. Civic associations create the repeated interactions that build trust, the shared experiences that generate norms of reciprocity, and the organizational competencies that enable more complex forms of collective action. For purpose ecosystems, the density and diversity of associational life in a territory is therefore a direct indicator of the relational substrate available for ecosystem formation. Territories with rich associational life have more potential ecosystem actors, more pre-existing relationships among those actors, and stronger norms of collaboration that reduce the barriers to multi-stakeholder engagement.

The capacity of a territory to mobilize and deploy resources for collective purposes is a distinct enabling condition not fully captured by either institutional quality or social capital. Resource mobilization capacity refers to the practical ability of territorial actors to access, coordinate, and deploy financial, human, and organizational resources in pursuit of shared goals. It encompasses the financial infrastructure (including grant-making institutions, impact investors, public funding mechanisms, and philanthropic organizations) that makes resources available for purpose ecosystem activities, as well as the organizational competencies needed to design, implement, and evaluate complex multi-actor programs, and the coordination mechanisms that enable diverse actors to pool their resources in ways that generate collective value greater than the sum of individual contributions.

Resource mobilization theory, developed by McCarthy and Zald (1977) and extended by Edwards and McCarthy (2004), provides important insights into how territorial actors assemble and deploy the resources needed for collective action. The key insight is that the emergence and success of collective action depends not only on the grievances or aspirations that motivate actors but on their capacity to access and organize the resources (money, labor, legitimacy, and organizational infrastructure) that sustained collective action requires. For purpose ecosystems, this means that territories with strong resource mobilization capacity (where diverse actors have demonstrated their ability to work together in deploying resources for shared purposes) are better positioned to generate and sustain purpose ecosystems than territories where resource mobilization is fragmented, contentious, or institutionally thin.

Absorptive capacity applies equally to the resource mobilization dimension of territorial enabling conditions. A territory may have significant financial resources available (through EU structural funds, national public investment, or private philanthropy) without having the organizational capacity to absorb and deploy those resources effectively. The implementation capacity of territorial actors (their ability to design credible programs, manage complex partnerships, monitor performance, and learn from experience) is as important as the volume of available funding. Territories with high implementation capacity can leverage relatively modest financial resources into significant collective impact. Territories with low implementation capacity may struggle to deploy even generous funding effectively.

The theoretical analysis of the preceding sections points toward a multi-dimensional understanding of territorial enabling conditions for purpose ecosystem emergence. No single condition is sufficient. Institutional quality without social capital generates administratively capable but relationally thin ecosystems. Social capital without institutional quality generates collaborative energy that cannot be channeled into effective collective action. Knowledge infrastructure without resource mobilization capacity generates ideas that cannot be implemented. Resource mobilization capacity without knowledge infrastructure generates activity without innovation. Effective territorial enabling conditions require the simultaneous presence and mutual reinforcement of all four dimensions.

This multi-dimensional, complementary understanding of enabling conditions provides the theoretical foundation for the Purpose Ecosystem Propensity Index (PEPI). It is a composite territorial indicator that captures the likelihood of purpose ecosystem emergence as a function of the joint presence and balance of structural and relational enabling conditions. The PEPI is not a measure of policy performance or a prescription for intervention. It is a diagnostic instrument for reading the territory. It is useful for understanding which territories have the latent conditions that make purpose ecosystem emergence more likely, where the most significant gaps are, and how the balance between structural and relational conditions varies across territorial contexts.

Obviously, the translation of enabling conditions into quantitative indicators requires inevitable compromises. Some conceptually relevant dimensions, such as the quality of informal relationships, resist full operationalization.

The PEPI rests on a foundational theoretical distinction between two qualitatively different but complementary types of enabling conditions that are structural and relational (see Table 2.1). Structural components are related to the formal, institutional, financial, and cognitive assets that provide the tangible framework within which ecosystem collaboration can occur. Relational infrastructure is related to social capital, trust networks, and civic cultures that provide the informal, relational substrate on which effective multi-actor coordination depends. Formal institutional quality and relational social capital are complementary, not substitutable, inputs to collective action (Coleman, 1990; Fukuyama, 1995; Ostrom, 1990; Putnam, 1993). Territories rich in structural but poor in relational foundation may possess the formal capacity for purpose ecosystem development without the trust, norms, and social networks needed to activate it. Territories rich in social capital but poor in institutional quality or resource capacity may possess strong relational foundations for cooperation without the structural scaffolding that gives it direction and scale. Effective ecosystem readiness requires both.

The hard structural component is operationalized through four complementary pillars, each capturing a distinct dimension of the structural enabling environment for purpose ecosystem emergence:

1. Ecosystem Activation Capacity. It measures the operational density of socially oriented organizations and initiatives in a territory. This is not a measure of policy activity but it is a measure of the vitality and organizational reach of the social sector. It is the degree to which there already exist organizations capable of designing and implementing coordinated multi-actor interventions. A territory with high ecosystem activation capacity has a rich organizational ecology of social enterprises, civic organizations, community groups, and mission-driven intermediaries that constitute the potential actor base for purpose ecosystem formation. The proxy used to operationalize this dimension (the number of social-oriented projects per 100,000 inhabitants drawn from cohesion policy databases) captures this organizational density indirectly.
2. Institutional Quality. It measures the administrative effectiveness, impartiality, and anti-corruption capacity of territorial governance institutions, operationalized through the European Quality of Government Index (EQI). Institutional quality is the meta-condition that enables all other structural capacities to be effectively deployed. High institutional quality reduces the transaction costs of multi-actor collaboration, protects against governance capture, and builds the public trust that enables voluntary compliance with collective rules. A territory with high institutional quality provides purpose ecosystems with the governance infrastructure they need to operate effectively. We refer to credible rules, impartial dispute resolution, and accountable public institutions that can be genuine partners rather than obstacles or predators.

Table 2.1 Components of the Purpose Ecosystem Propensity Index (PEPI)

PEPI Component	Dimension	What It Captures	Proxy Indicator
Structural component	Ecosystem Activation Capacity	Organizational density and reach of socially oriented actors in the territory	Numbers of social projects per 100,000 inhabitants
Structural component	Institutional Quality	Administrative effectiveness, impartiality, and anti-corruption capacity	European Quality of Government Index (EQI)
Structural component	Resource Mobilization Capacity	Demonstrated ability to deploy resources for collective social purposes	Public expenditure on social projects per capita
Structural component	Knowledge Infrastructure	Cognitive and innovation capacity of the territory	R&D expenditure as % of regional GDP
Relational component	Generalized Trust	Willingness to cooperate with unknown actors	Trust survey indicators (BES)
Relational component	Voluntary Activity	Density of associational life and civic organizations	Voluntary participation rates (BES)
Relational component	Social Participation	Active engagement in collective processes	Social participation rates (BES)
Relational component	Civic Engagement	Participation in civic and political life	Civic and political participation rates (BES)

Source Authors' own elaboration

3. Resource Mobilization Capacity. It measures the territory's demonstrated ability to translate resources into coordinated collective action. The proxy used is public expenditures allocated to socially relevant projects per capita. It captures not the generosity of public funding but the practical capacity of the territory to absorb, manage, and deploy resources for collective social purposes. A territory with high resource mobilization capacity has organizations with the competencies, networks, and institutional relationships needed to assemble multi-actor coalitions around shared objectives and deliver complex, multi-dimensional programs. This is the organizational infrastructure that makes purpose ecosystem activities viable. Without it, even well-intentioned actors with strong social capital and good ideas cannot sustain the operational demands of purpose ecosystem development.
4. Knowledge Infrastructure. It measures the cognitive and innovation capacity of the territory through regional expenditure in research and development as a percentage of regional GDP. Purpose ecosystems address complex challenges

that require sophisticated analytical capabilities and continuous innovation. The knowledge infrastructure of a territory determines whether these capabilities are available within the ecosystem's geographic reach, and whether the ecosystem can connect to the global knowledge frontier in the domains relevant to its Transformative Purpose. A territory with strong knowledge infrastructure can generate purpose ecosystems that draw on the best available scientific and technical knowledge, produce innovations of sufficient quality to address systemic challenges, and attract the knowledge workers and research institutions that amplify ecosystem impact.

The relational component of the PEPI is operationalized through the Social Capital Index (SCI), constructed from four dimensions of social capital that correspond to the bonding, bridging, and linking forms identified in the foundational literature. They are generalized trust, participation in voluntary activities, social participation, and civic and political participation. Generalized trust represents the most fundamental relational precondition for purpose ecosystem development. It is the willingness to cooperate with actors whose intentions and capabilities are not fully known, which is essential in multi-stakeholder systems where formal contracts cannot specify all relevant contingencies. Participation in voluntary activities captures the density of associational life through which social capital is produced and reproduced. Social participation and civic and political participation capture the degree to which territorial populations are actively engaged in collective processes, the habits of collaborative deliberation and shared decision-making that purpose ecosystem governance depends on.

The integrated PEPI is computed by combining the hard structural and relational components through an aggregation method that operationalizes the theoretical claim that are jointly necessary for ecosystem readiness. The specific aggregation method adopted, and its mathematical formalization, are presented in Sect. 2.3, which also reports the empirical results for the 20 Italian regions. Neither component is sufficient alone. Both must be simultaneously present for a territory to exhibit high propensity for purpose ecosystem emergence. Regions with strong structural endowments but weak relational density may possess the formal capacity for coordinated action without the trust and civic culture needed to activate it organically. Regions with dense social capital but insufficient institutional or resource capacity may exhibit latent cooperative potential that cannot be realized at the scale required by purpose ecosystems. High PEPI scores therefore represent genuinely balanced territorial development across structural and relational dimensions.

The PEPI provides a dual classification of territories that is more informative than a simple ranking.

At the scalar level, five ecosystem readiness levels are defined: Fragile Ecosystem Context (L1), where insufficient enabling conditions exist across multiple structural and relational pillars; Emerging Ecosystem (L2), where partial structural development exists but remains unstable; Developing Ecosystem (L3), where growing coherence appears among pillars but without sustained transformative capacity; Consolidated Ecosystem (L4), where balanced infrastructure supports demonstrated multi-actor coordination; and Advanced Purpose Ecosystem (L5), where strong and mutually reinforcing hard structural and relational capacities provide optimal conditions for purpose ecosystem emergence.

The configurational typology complements the scalar classification by crossing the hard structural and relational dimensions to identify four territorial profiles. Structurally Fragile Contexts (with low hard structural and low relational components) face the most severe constraints on organic purpose ecosystem emergence. Relationally Endowed but Structurally Weak Contexts (with strong social capital but insufficient institutional, financial, or knowledge infrastructure) represent territories where latent cooperative potential exists but cannot be realized without structural development. Technocratically Equipped Contexts (with strong structural but weak relational embeddedness) risk generating administratively driven but community-shallow configurations. Fully Embedded Purpose Ecosystems (with strong structural and relational components) represent the optimal conditions for organic purpose ecosystem emergence.

It is essential to interpret the PEPI correctly. It measures propensity, not destiny. A territory with a low PEPI is not incapable of generating purpose ecosystems. It simply faces greater structural and relational challenges that make organic emergence less likely and less sustainable without additional support. A territory with a high PEPI is not guaranteed to generate purpose ecosystems. The index captures enabling conditions, not the shared values and purpose vision that are the ultimate origin of every purpose ecosystem. What the PEPI provides is a diagnostic map of the territorial landscape, an evidence-based picture of where the soil is most fertile for purpose ecosystem emergence, where the most significant gaps are, and what kind of support would most effectively address those gaps. Section 2.3 applies this diagnostic logic to the Italian context, showing how the 20 regions differ not only in their aggregate PEPI scores but in the specific structural and relational profiles through which they achieve those scores.

Box 2.1 A Focus on PEPI

The PEPI tells us how fertile a territory is for organic purpose ecosystem emergence; which enabling conditions are strong and which are weak; whether the gap is primarily structural (institutions, resources, knowledge) or relational (trust, civic participation, associational life); how a territory compares with others in its enabling condition profile.

A high PEPI score means the territory has fertile conditions for organic emergence. A low PEPI score means the territory needs more support for ecosystems to develop. But in neither case does the PEPI substitute for the shared values, the genuine purpose, and the collaborative relationships that are the true origin of every purpose ecosystem.

2.2 Institutional Intervention for Purpose Ecosystem Development

Institutional intervention, through national programs, European structural funds, regional development strategies, or targeted ecosystem policies, can create structural preconditions for purpose ecosystem development. It can build institutional capacity, mobilize financial resources, create legitimacy for multi-actor collaboration, lower the transaction costs of partnership formation, and provide the organizational infrastructure that sustains ecosystem activities over time. What it cannot do is generate the shared values, the emergent purpose, and the authentic relational trust that are the constitutive elements of a genuine purpose ecosystem. Purpose cannot be assigned by a policy document, installed by a governance structure, or funded into existence by a grant program. It must emerge from within, from actors who genuinely share a vision of the social and environmental change they want to create.

This distinction between structural preconditions and purpose emergence has practical implications for how institutional interventions should be designed, implemented, and evaluated. Interventions that focus exclusively on structural conditions without attending to the relational and cultural conditions that enable authentic purpose to emerge, risk producing hollow ecosystems. These can be seen as formally correct organizational configurations that satisfy the criteria of funders and evaluators without generating the genuine collaborative energy and transformative commitment that distinguish purpose ecosystems from funded networks. This risk is not hypothetical. It is a well-documented phenomenon in the social innovation literature, where it appears under various names including purpose-washing, mission drift, and institutional isomorphism.

Understanding the conditions under which institutional intervention can genuinely support purpose ecosystem development is relevant. The management and public administration literatures offer several frameworks.

One of the most direct frameworks for understanding how institutional intervention can support purpose ecosystem formation is transaction cost theory, developed by Coase (1937) and Williamson (1985) and extended to the governance of multi-actor collaborations by Ansell and Gash (2008) and Provan and Kenis (2008). Purpose ecosystems face significant transaction costs such as the costs of aggregating potential partners, assessing their values and capabilities, negotiating collaborative arrangements or resolving disputes when they arise. In territories with weak institutional infrastructure, high social fragmentation, and low cross-sector collaboration, these transaction costs can be a limit. They can prevent purpose ecosystems from forming even when the conditions for purpose action are present. Institutional intervention can reduce transaction costs by providing infrastructure that brings potential partners together, creating shared standards and frameworks that reduce the costs of negotiating, offering legal and organizational support, and building the relational capital that reduces the costs of trust development over time.

Also, institutional entrepreneurship theory (e.g., Battilana et al., 2009; DiMaggio, 1988; Garud et al., 2007) provides a useful framework to better understand the creation of institutional conditions in a territory. It explains how specific actors can play a catalytic role in creating the institutional conditions for new organizational forms to emerge. Institutional entrepreneurs are actors who can create or transform institutions in ways that enable new forms of collective action. In the context of purpose ecosystem, institutional entrepreneurs might include public agencies that provide funding and organizational support for nascent ecosystem initiatives, universities that offer convening space and research expertise to multi-actor collaborations (Cunningham et al., 2022; Del Giudice et al., 2016; Nicotra et al., 2021), finance institutions that provide patient capital for ecosystem activities, or civic organizations. In this theory, institutional conditions are not purely structural determinants of what is possible. They are actively shaped by specific actors who can create new opportunities.

For institutions aimed at supporting purpose ecosystem development, interventions should create spaces in which diverse actors can develop shared understandings of the challenges of their territory to enable authentic purpose. The governance conditions under which multi-actor collaborations can sustain genuine collaborative engagement emerge from the collaborative governance theory, developed by Ansell and Gash (2008), Emerson et al. (2012), and Ostrom (1990, 2010). Collaborative governance requires face-to-face dialogue that builds trust and shared understanding, commitment, and institutional arrangements that give all participants genuine voice and influence over collective decisions.

Besides, institutional interventions should build the inter-organizational relationships that purpose ecosystems require. Relational capital is the relevant capital needed for a purpose ecosystem. Relational capital refers to the trust, respect, and friendship that develop between alliance partners through repeated interaction, shared experience, and demonstrated reciprocity (Dyer & Singh, 1998; Kale et al., 2000; Lavie, 2006). Building relational capital takes time and cannot be rushed by formal agreements or incentive structures. But institutional interventions can create the conditions that enable relational capital to develop by bringing together actors who share values but lack prior relationship, by providing structures for repeated interaction, by creating shared projects that require genuine collaboration and generate shared memories of success and failure, and by building the organizational competencies (including boundary-spanning roles, knowledge-sharing routines, and conflict resolution mechanisms) that sustain relationships through difficulties.

The literature on mission-oriented innovation policy, developed by Kattel and Mazzucato (2018), Mazzucato (2018), and Schot and Steinmueller (2018), provides important guidance on how public institutions can create enabling conditions for systemic transformation without prescribing its direction or content. Mission-oriented policy sets ambitious societal goals (such as climate neutrality, zero poverty, or universal access to quality education) without specifying the technological or organizational pathways through which those goals will be achieved. By creating a clear directional signal without prescribing the solution, mission-oriented policy creates space for diverse actors to innovate around a shared challenge, generating the kind of purpose, multi-actor experimentation that characterizes genuine purpose ecosystem development. The European Green Deal provides a prominent example of mission-oriented framing that can support purpose ecosystem emergence without determining its form.

The relationship between the PEPI and institutional intervention follows directly from this analysis (Table 2.2). The PEPI provides the diagnostic foundation for differentiated intervention strategies tailored to the specific enabling condition profiles of different territories. Different PEPI profiles imply different intervention priorities and different risks of hollow ecosystem formation.

In Structurally Fragile Contexts (territories with low hard structural and low relational components), institutional intervention faces the greatest challenge and the highest risk of hollow ecosystem formation. These territories lack the organizational infrastructure to implement complex multi-actor programs and the relational capital to sustain genuine collaborative engagement. Interventions in these contexts should be addressed to long-term capacity building, investing in institutional quality, building associational life, developing organizational competencies, and creating the conditions for trust and social capital to develop over time. Attempting to create formal purpose ecosystem structures before the enabling conditions exist risks generating empty shells without genuine purpose.

In Relationally Endowed but Structurally Weak contexts, institutional intervention can be highly effective if it focuses on structural gap. These territories have the relational foundations for genuine collaborative engagement, that are trust, civic participation, and associational life. However, they lack institutional quality,

Table 2.2 PEPI profiles and differentiated policy strategies

PEPI Profile	Primary Challenge	Policy Priority	Main Risk
Structurally Fragile (low structural, low relational)	Absence of organizational and relational foundations	Long-term capacity building: institutions, associations, competencies	Hollow ecosystems consuming resources without genuine purpose
Relationally Endowed but Structurally Weak (low structural, high relational)	Latent collaborative potential not realizable at scale	Structural gap-filling: institutional quality, resources, knowledge	Relational energy dissipated without organizational scaffolding
Technocratically Equipped (high structural, low relational)	Organizational capacity without community rootedness	Trust-building and community engagement: participatory governance	Community-shallow ecosystems lacking authentic purpose
Fully Embedded (high structural, high relational)	Scaling and sustaining organic emergence	Catalytic amplification: seed funding, regulatory support, networking	Policy over-engineering that crowds out organic emergence

Source Authors' own elaboration

resource mobilization capacity, and knowledge infrastructure needed to channel that relational energy into sustained purpose ecosystem activity. Targeted interventions should provide organizational support and financial resources, activating the latent collaborative potential of these territories.

Technocratically Equipped Contexts are territories with the organizational competencies and resource capacity to implement complex multi-actor programs, but lack the social capital and civic engagement needed for genuine community co-creation. In this case, the aim of interventions should be trust-building, community engagement, and the development of participatory governance mechanisms. The most important contribution that policy can make in these contexts is creating genuine spaces for community voice and influence, ensuring that purpose ecosystem activities are co-designed with the communities they want to serve.

In Fully Embedded Purpose Ecosystems (territories with strong structural and relational infrastructure), institutional intervention can play a catalytic rather than a constitutive role. In this case, policy intervention is most effective when it amplifies and accelerates processes that are already underway. This means providing seed funding for nascent ecosystem initiatives, supporting new organizational forms, connecting local ecosystems with regional and national networks, or investing in the shared digital and physical infrastructure that enables ecosystem activities to scale.

Additionality, a concept widely used in the evaluation of public investment programs, provides an important evaluative criterion for institutional interventions in support of purpose ecosystem development. It asks whether the intervention generated activities, relationships, or outcomes that would not have occurred in its absence. For purpose ecosystem support interventions, the most important form of additionality is not financial (whether the intervention funded activities that would have been funded anyway) but relational and purpose. Intervention should be able to generate genuine collaborative relationships, authentic shared purpose, and organizational capabilities that would not have developed without institutional support. Evaluating purpose ecosystem support interventions against this relational and purpose additionality standard (rather than merely against output metrics) is essential for distinguishing interventions that genuinely support purpose ecosystem emergence from those that merely simulate it.

The timing of institutional intervention matters enormously for its effectiveness and its risk of crowding out organic emergence. Interventions that arrive too early (before actors have developed the shared values and relational trust that constitute authentic purpose) may impose governance structures and reporting requirements that constrain rather than enable the organic processes through which purpose emerges. Interventions that arrive too late (after purpose ecosystems have already developed their governance arrangements and relational networks) may add bureaucratic overhead without contributing to ecosystem capacity. The most effective timing for institutional intervention is the moment when a nascent ecosystem has developed sufficient shared purpose and relational capital to sustain genuine collaborative engagement, but lacks the structural resources, organizational capacity, or institutional legitimacy to scale its activities to the level required for transformative impact.

At this moment, institutional intervention can provide the structural scaffolding that enables the ecosystem to grow without distorting the authentic purpose that drives it.

The governance of policy-initiated ecosystems deserves specific attention because it is at the governance level that the tension between institutional structure and authentic purpose is most acutely felt. When institutions create formal governance structures for purpose ecosystems, they inevitably introduce elements of hierarchical authority, formal accountability, and procedural compliance that can conflict with the distributed agency, participatory legitimacy, and adaptive governance that authentic purpose ecosystems require. Managing this tension requires careful attention to governance design. This means creating accountability mechanisms that are transparent and credible; developing evaluation frameworks that capture the relational and purpose dimensions of ecosystem performance alongside the output metrics that funders require; and building the institutional cultures within funding and oversight bodies that recognize the emergent, bottom-up character of genuine purpose and resist the temptation to prescribe or control it.

Finally, institutional intervention in support of purpose ecosystem development must attend to the long-term sustainability of the ecosystems it helps create. Purpose ecosystems that depend indefinitely on external funding and institutional support are not purpose ecosystems in the full sense. They are institutionally sustained networks whose collaborative activity would cease if external support were withdrawn. The goal of institutional intervention should therefore be not merely to create purpose ecosystems but to strengthen the enabling conditions that make them self-sustaining. This long-term, enabling-condition-building orientation distinguishes genuinely developmental institutional intervention from short-term ecosystem simulation.

Box 2.2 What Policy Can and Cannot Do for Purpose Ecosystems

Policy CAN: Lower the transaction costs of multi-stakeholder collaboration by providing convening infrastructure, shared frameworks, and organizational support. Mobilize financial resources that individual actors cannot assemble independently. Create institutional legitimacy for new forms of multi-actor collaboration. Fill structural gaps in territories with strong relational foundations but weak institutional or resource capacity. Catalyze and amplify organic emergence in territories where purpose ecosystems are already forming. Build the enabling conditions that make future organic emergence more likely.

Policy CANNOT: Generate the shared values that motivate genuine purpose ecosystem actors. Create the relational trust that sustains authentic multi-actor collaboration. Prescribe the direction or content of purpose because this must emerge from the actors and communities involved. Substitute for the community co-creation that gives purpose ecosystems their legitimacy. Guarantee that formal ecosystem structures will generate genuine transformative impact. Prevent purpose-washing in ecosystems whose governance is captured by dominant actors.

> The fundamental principle is that policy creates the container. Purpose fills it from within. The quality of the container determines whether purpose can emerge, develop, and sustain itself. But the container alone is not enough.

2.3 A Benefit-of-the-Doubt Approach Applied to Italian Regions

The theoretical framework developed in Sects. 2.1 and 2.2 raises a practical question. The PEPI has been defined conceptually, its eight components have been identified, and its configurational typology has been articulated. But how should those eight components be combined into a single composite score? And what weights should each component receive? These are not trivial choices. Different weighting schemes can produce very different territorial rankings, and the choice of aggregation method embeds assumptions about the relative importance of different enabling conditions that the theory alone cannot resolve.

The standard approach in composite indicator construction of assigning equal or arbitrary weights to each component is unsatisfying. It assumes that all enabling conditions contribute equally to purpose ecosystem propensity, which the theoretical analysis of Sect. 2.1 does not support. The four structural pillars and the four relational dimensions have different empirical distributions and different causal pathways to ecosystem emergence. Besides, equal weighting ignores the heterogeneity of territories. As the configurational typology of Sect. 2.1 makes clear, different territories have different enabling condition profiles. Some are structurally strong but relationally thin; others are rich in social capital but institutionally fragile. A single fixed set of weights imposes a uniform standard of evaluation that erases exactly the territorial differentiation that the PEPI is designed to reveal. The Benefit-of-the-Doubt approach adopted in this section addresses both concerns directly. By allowing each territory to be evaluated under the most favorable weighting scheme, the method identifies the dimensions in which each territory performs relatively better and reflects those strengths in the composite score. The result is a ranking that is not only more statistically defensible than equal weighting alternatives but is also richer in diagnostic content. The weight vectors reported for each region (Table 2.3) reveal the structural and relational profile through which each territory achieves its overall PEPI score, providing the fine-grained territorial intelligence that the differentiated policy strategies of Sect. 2.2 require.

Table 2.3 PEPI scores and weight vectors for the 20 Italian regions

No.	Region	w_1	w_2	w_3	w_4	w_5	w_6	w_7	w_8	Score
1	Liguria	0.07	0.2	0.3	0.21	0.06	0.11	0.1	0.56	1
2	Trentino-South Tyrol	0.07	0.19	0.3	0.21	0.06	0.12	0.1	0.56	1
3	Emilia-Romagna	0.07	0.1	0.35	0.21	0.06	0.08	0.21	0.56	0.99
4	Sardinia	0.03	0.2	0.5	0.1	0.06	0.12	0.21	0.39	0.98
5	Friuli-Venezia Giulia	0.07	0.2	0.3	0.21	0.06	0.11	0.1	0.56	0.96
6	Tuscany	0.03	0.1	0.39	0.21	0.06	0.07	0.21	0.56	0.95
7	Marche	0.03	0.2	0.5	0.1	0.06	0.12	0.21	0.39	0.93
8	Piedmont	0.07	0.2	0.29	0.21	0.06	0.06	0.17	0.56	0.9
9	Veneto	0.03	0.13	0.38	0.21	0.06	0.06	0.21	0.56	0.9
10	Aosta Valley	0.03	0.2	0.39	0.1	0.12	0.08	0.1	0.56	0.9
11	Lazio	0.07	0.1	0.34	0.21	0.08	0.06	0.21	0.56	0.85
12	Umbria	0.03	0.1	0.46	0.1	0.06	0.1	0.21	0.56	0.85
13	Lombardy	0.03	0.2	0.34	0.21	0.06	0.06	0.16	0.56	0.83
14	Apulia	0.03	0.1	0.5	0.1	0.06	0.12	0.21	0.5	0.8
15	Abruzzo	0.03	0.1	0.39	0.21	0.06	0.07	0.21	0.56	0.77
16	Basilicata	0.07	0.2	0.5	0.1	0.11	0.12	0.21	0.28	0.71
17	Campania	0.03	0.2	0.5	0.21	0.12	0.12	0.1	0.3	0.7
18	Molise	0.07	0.1	0.5	0.21	0.12	0.12	0.21	0.28	0.58
19	Sicily	0.07	0.1	0.5	0.21	0.12	0.12	0.21	0.28	0.58
20	Calabria	0.03	0.2	0.5	0.21	0.12	0.12	0.1	0.3	0.57

Source Authors' own elaboration

The empirical application focuses on the 20 Italian regions, using the eight indicators introduced in Table 2.1 and so the four hard structural pillars and the four relational dimensions of the Social Capital Index. The use of a multi-criteria composite indicator to benchmark Italian territorial ecosystems builds on a growing tradition of empirical ecosystem assessment (Corrente et al., 2019; Nicotra et al., 2018), which has demonstrated the value of data-driven approaches for comparing territorial enabling conditions across heterogeneous regions.

The final composite indicator, which measures the performance of each region, is obtained by aggregating the eight different indicators. The literature proposes several approaches for determining the weights assigned to each indicator and the corresponding aggregation method (Greco et al., 2019a). In this study, we adopt the Data Envelopment Analysis (DEA) method (Charnes et al., 1978), in which weights are endogenously determined from the data, thereby reducing subjectivity in the weighting process. More specifically, DEA relies on linear programming models to evaluate the performance of decision-making units (e.g., countries, universities, or regions). The final score assigned to each unit is computed as the ratio between a weighted sum of outputs and a weighted sum of inputs. A specific approach within this framework is the Benefit-of-the-Doubt (BoD) method (Cherchye et al., 2007), which considers a single input, the composite indicator, and multiple outputs corresponding to its components. This approach allows each unit to be evaluated under the most favorable set of weights, subject to a set of common constraints. Many authors have used the BoD to construct composite indicator in different context, such as Technology Achievement Index (Cherchye et al., 2008a, 2008b), Human Development Index (Cherchye et al., 2008b), Health Care Quality (Shwartz et al., 2009), Teacher Effectiveness Scores (Rogge, 2011), Digital Access Indicator (Gaaloul & Khalfallah, 2014), and the Gender Gap in the Energy Sector (Henriques et al., 2024).

Let $A = \{a_1, a_2, \ldots, a_{20}\}$ denote the set of the 20 Italian regions to be evaluated across eight criteria $G = \{g_1, g_2, \ldots, g_8\}$. Each criterion $g_i \in G$ is defined as a real value function $g_i : A \rightarrow \mathbb{R}$, where $g_i(a_j)$ represents the evaluation of region a_j on criterion g_i.

The BoD method provides a vector of weights $w_{1,j}, \ldots, w_{8,j}$ for each region a_j. All criteria are assumed to be monotonically increasing and are normalized following the procedure described in Greco et al. (2019b) (Tables 2.4 and 2.5).

For a given region a_c, the associate composite indicator $CI(a_c)$, is obtained by solving the following linear programming (LP) problem.

Equation 1—Benefit-of-the-Doubt LP problem first step

$$CI(a_c) = \max \sum_{i=1}^{8} w_{c,i} \cdot g_i(a_c), \textit{ subject to,} \tag{2.1}$$

$$\left.\begin{array}{ll} \sum_{i=1}^{8} w_{j,i} \cdot g_i(a_j) \leq 1, & \textit{for all } j = 1, \ldots, 20 \\ w_{c,i} \geq 0, & \textit{for all } i = 1, \ldots, 8 \end{array}\right\}, \tag{2.2}$$

where

- The objective function (2.1) is the computation of the composite indicator for the considered region. Its aim is to maximize its score, that means the highest evaluation relative to all the other regions;
- The first constraint of (2.2) ensures that no region can achieve a score greater than one, thus acting as a normalization condition;
- The second constraint of (2.2) prevents the assignment of negative weights.

Table 2.4 Raw indicator values for the 20 Italian regions

Region	No. of projects (2021–2027) per 100,000 inhabitants	EQI (2024)	Public expenditures or payments allocated to inclusion-related projects per capita (2021–2027)	Regional expenditure R&D per GDP (2023)	Generalized trust (% 2024)	Participation in voluntary activities (% 2024)	Social participation (% 2024)	Civic and political participation (% 2024)
Abruzzo	12.97	−0.903	117.69 €	0.011	0.178	0.08	0.308	0.588
Basilicata	25.30	−1.079	252.23 €	0.006	0.188	0.081	0.254	0.521
Calabria	9.04	−1.278	121.11 €	0.006	0.225	0.06	0.199	0.522
Campania	8.16	−1.215	278.83 €	0.012	0.198	0.045	0.21	0.519
Emilia-Romagna	29.32	−1.102	72.14 €	0.021	0.234	0.107	0.321	0.666
Friuli-Venezia Giulia	25.71	0.718	36.89 €	0.017	0.253	0.12	0.316	0.628
Lazio	10.49	−1.068	38.79 €	0.018	0.24	0.077	0.312	0.627
Liguria	48.15	0.081	318.79 €	0.014	0.263	0.075	0.279	0.625
Lombardy	3.22	−0.225	34.92 €	0.012	0.243	0.097	0.315	0.62
Marche	15.50	−0.152	310.97 €	0.009	0.219	0.088	0.29	0.605
Molise	29.93	−1.938	57.77 €	0.008	0.194	0.072	0.288	0.515
Piedmont	24.35	−0.28	98.51 €	0.021	0.239	0.095	0.302	0.603
Apulia	13.28	−1.266	336.57 €	0.008	0.193	0.059	0.274	0.553
Sardinia	11.60	−0.225	416.34 €	0.008	0.222	0.086	0.27	0.602
Sicily	6.13	−2.059	247.32 €	0.010	0.157	0.044	0.223	0.461
Tuscany	22.04	−0.514	173.97 €	0.014	0.207	0.089	0.299	0.656
Trentino-South Tyrol	3.16	0.244	35.11 €	0.011	0.353	0.171	0.392	0.657
Umbria	11.09	−0.945	66.84 €	0.009	0.232	0.11	0.323	0.643

(continued)

Table 2.4 (continued)

Region	No. of projects (2021–2027) per 100,000 inhabitants	EQI (2024)	Public expenditures or payments allocated to inclusion-related projects per capita (2021–2027)	Regional expenditure R&D per GDP (2023)	Generalized trust (% 2024)	Participation in voluntary activities (% 2024)	Social participation (% 2024)	Civic and political participation (% 2024)
Aosta Valley	6.50	−0.127	127.60 €	0.005	0.294	0.109	0.298	0.642
Veneto	2.89	−0.403	56.21 €	0.012	0.253	0.107	0.333	0.65

Source Authors' own elaboration

Table 2.5 Normalized indicator values for the 20 Italian regions

Region	No. of projects (2021–2027) per 100,000 inhabitants	EQI (2024)	Public expenditures or payments allocated to inclusion-related projects per capita (2021–2027)	Regional expenditure R&D per GDP (2023)	Generalized trust (% 2024)	Participation in voluntary activities (% 2024)	Social participation (% 2024)	Civic and political participation (% 2024)
Abruzzo	0.456	0.448	0.441	0.475	0.299	0.449	0.568	0.479
Basilicata	0.637	0.406	0.63	0.303	0.338	0.455	0.361	0.286
Calabria	0.399	0.359	0.446	0.306	0.483	0.33	0.151	0.289
Campania	0.386	0.374	0.667	0.514	0.377	0.241	0.193	0.28
Emilia-Romagna	0.696	0.401	0.377	0.845	0.519	0.609	0.617	0.704
Friuli-Venezia Giulia	0.643	0.836	0.327	0.695	0.593	0.686	0.598	0.595
Lazio	0.42	0.409	0.33	0.738	0.542	0.431	0.583	0.592
Liguria	0.973	0.683	0.723	0.606	0.633	0.419	0.457	0.586
Lombardy	0.313	0.61	0.325	0.509	0.554	0.55	0.595	0.572
Marche	0.493	0.628	0.712	0.417	0.46	0.496	0.499	0.528
Molise	0.705	0.201	0.357	0.358	0.361	0.401	0.491	0.269
Piedmont	0.623	0.597	0.414	0.842	0.538	0.538	0.545	0.523
Apulia	0.461	0.362	0.748	0.372	0.358	0.324	0.438	0.378
Sardinia	0.436	0.61	0.86	0.377	0.472	0.485	0.422	0.52
Sicily	0.356	0.172	0.623	0.433	0.216	0.235	0.242	0.113
Tuscany	0.589	0.541	0.52	0.576	0.413	0.502	0.533	0.676
Trentino-South Tyrol	0.312	0.723	0.325	0.46	0.986	0.989	0.889	0.678
Umbria	0.429	0.438	0.369	0.395	0.511	0.627	0.625	0.638

(continued)

Table 2.5 (continued)

Region	No. of projects (2021–2027) per 100,000 inhabitants	EQI (2024)	Public expenditures or payments allocated to inclusion-related projects per capita (2021–2027)	Regional expenditure R&D per GDP (2023)	Generalized trust (% 2024)	Participation in voluntary activities (% 2024)	Social participation (% 2024)	Civic and political participation (% 2024)
Aosta Valley	0.361	0.634	0.455	0.266	0.755	0.621	0.529	0.635
Veneto	0.308	0.568	0.354	0.512	0.593	0.609	0.663	0.658

Source Authors' own elaboration

A key feature of the model is that it provides a region-specific vector of weights, revealing the dimensions in which each region performs relatively better. This property enables a more distinct interpretation of performance beyond the aggregate score.

However, the solution of Equation 1 may assign zero weights to some indicators, implying that certain dimensions are effectively not considered in the evaluation of a given region. To address this issue, we adopt the procedure proposed by Roll and Golany (1993), which consists of the following steps:

1. Solve the BoD problem (Equation 1) to obtain the weights $w_{j,i}$. Then, compute the mean of weights ($\overline{w}_i$) for each criterion g_i;
2. Solve a new LP problem by adding to the set of constraints of Equation 1 the following one:

$$\frac{2 \cdot \overline{w}_i}{1+d} \leq w_{c,i} \leq \frac{2 \cdot d \cdot \overline{w}_i}{1+d}, \textit{for all } i = 1, \ldots, 8 \tag{2.3}$$

with $d \geq 1$.

The parameter d controls the admissible deviation of weights from their mean value. The bounds are constructed symmetrically around the average weight, while d determines the width of the interval and the degree of flexibility allowed. In case $d = 1$ the weights have no variability across their mean, while if d increases, the model tends to the unbounded BoD.

As the empirical evidence shown in Pedraja-Chaparro et al. (1997) and Gaaloul and Khalfallah (2014) suggests, we set $d = 2$. In this case, the constraint (2.3) becomes:

$$\frac{2 \cdot \overline{w}_i}{3} \leq w_{c,i} \leq \frac{4 \cdot \overline{w}_i}{3}$$

which means allowing each weight to vary within ±33% of its mean value, preventing extreme or zero weights while still allowing regions to reflect their relative strengths across dimensions.

The BoD model allows for the assessment of regions relative to one another while simultaneously identifying the relative importance assigned to the different dimensions of the purpose ecosystem.

Table 2.3 reports the overall score of each region (last column) and the corresponding weights, which reflect the relative specialization of each region across the eight dimensions. This dual information is particularly valuable, as it enables the identification of not only which regions perform better, but also how such performance is achieved.

A comparison between Emilia-Romagna and Lazio illustrates this point (Fig. 2.1). Even if the two regions show similar weights, they are ranked at the 3rd and the 11th position, respectively. This difference is not only driven by the weighting scheme itself, but rather by the underlying performance levels across the selected indicators. In other words, similar strategic emphasis across dimensions does not necessarily translate into comparable outcomes. More specifically, the difference in ranking is therefore almost entirely driven by the underlying performance levels: Emilia-Romagna outperforms Lazio across most of the indicators.

The model also enables meaningful comparisons between regions with similar overall performance. For example, both Liguria and Emilia-Romagna achieve almost the maximum score, but their underlying efficiency profiles differ substantially. This indicates that similar aggregate performance can be attained through different configurations of strengths across dimensions. In particular, Liguria, compared to Emilia-Romagna, places greater emphasis on Institutional quality, while vice versa, Emilia-Romagna assigns relatively higher importance to Social Participation.

As it is possible to see in Fig. 2.2, the results confirm several of the theoretical expectations developed in Sect. 2.1. Northern and central regions with historically strong institutional quality and social capital (Liguria, Trentino-South Tyrol, Emilia-Romagna, Friuli-Venezia Giulia) consistently achieve the highest PEPI scores. Southern regions (Molise, Calabria, Sicily, Campania) face a more challenging enabling condition profile across structural and relational dimensions, consistent with the long-documented North-South divide in Italian civic and institutional development. Crucially, however, the weight vectors reveal that similar aggregate scores can be achieved through different configurations of strengths, as the comparison between Liguria and Emilia-Romagna illustrates. This finding directly validates the configurational logic according to which territorial propensity for purpose ecosystem emergence is not a single-dimensional phenomenon but a multi-dimensional balance whose composition matters as much as its level. The practical implication is that territories with similar overall PEPI scores may require very different policy interventions depending on whether their enabling condition profile is structurally strong but relationally thin, relationally rich but institutionally fragile, or balanced across the two dimensions.

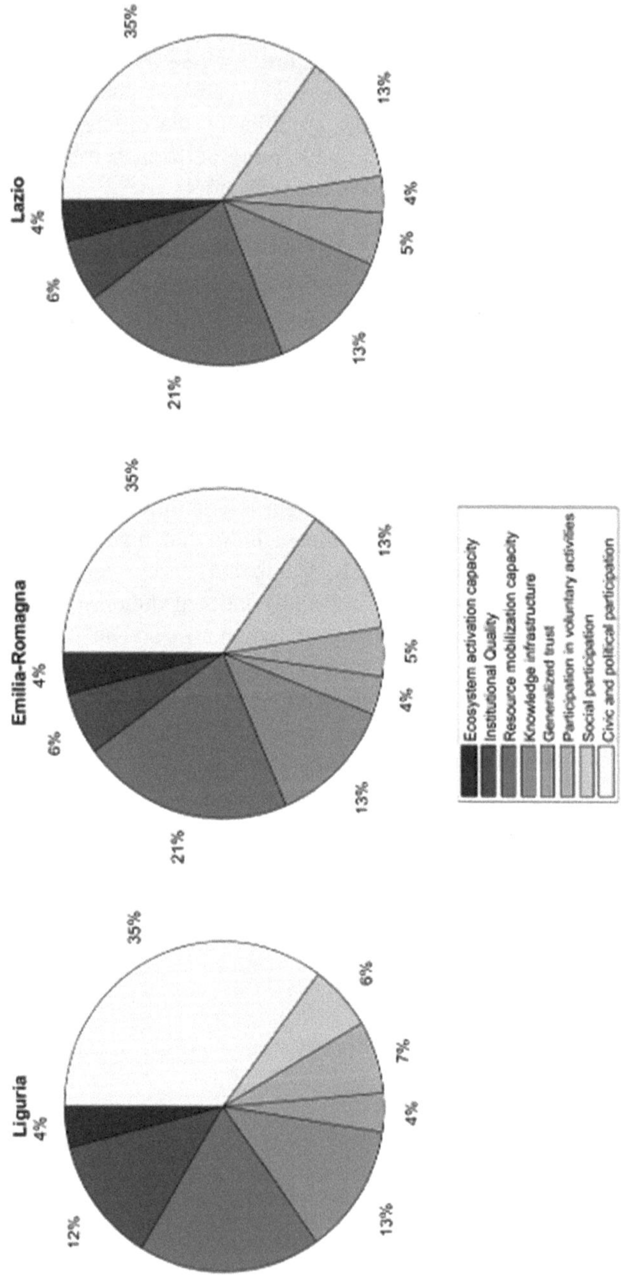

Fig. 2.1 Weight profiles for selected regions (Liguria, Emilia-Romagna, and Lazio). *Source* Authors' own elaboration

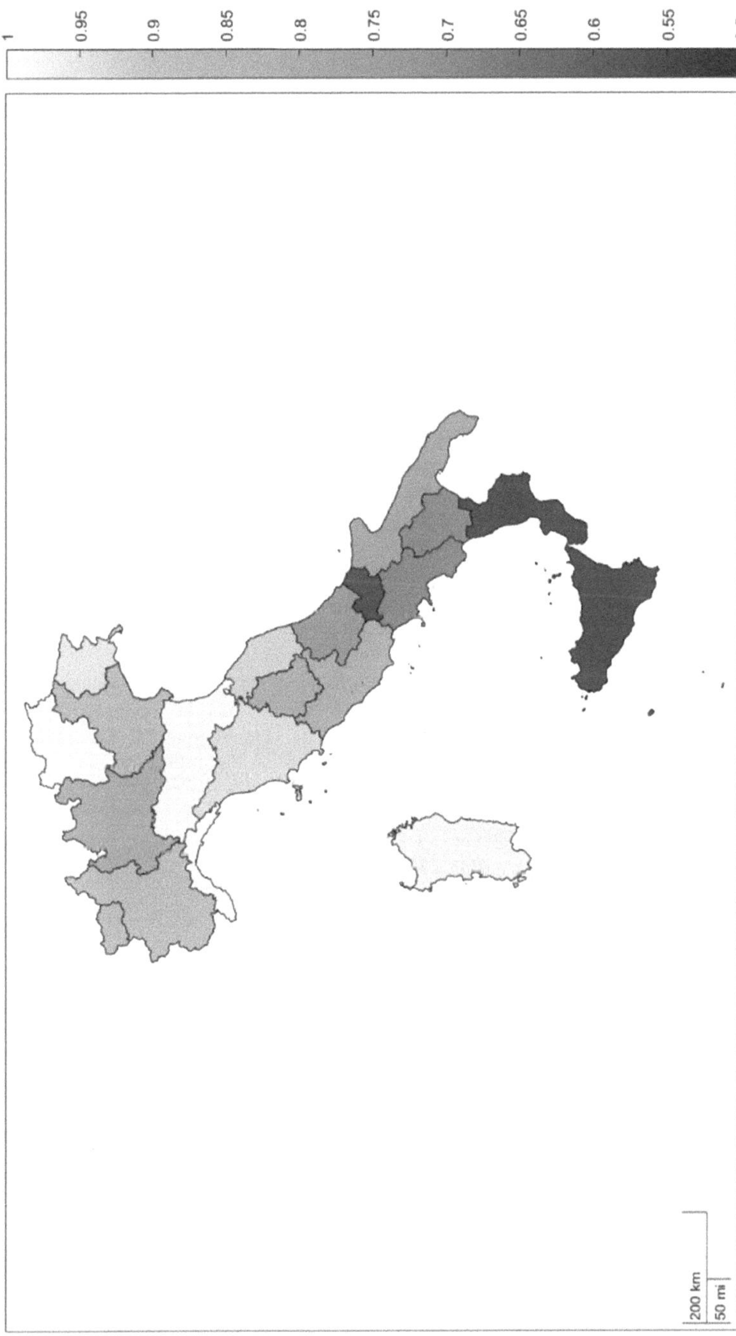

Fig. 2.2 Italian regions ranking according to the overall score in Table 2.3. *Source* Authors' own elaboration

References

Acemoglu, D., Johnson, S., & Robinson, J. A. (2001). The colonial origins of comparative development: An empirical investigation. *American Economic Review, 91*(5), 1369–1401.

Amable, B. (2003). *The diversity of modern capitalism.* Oxford University Press.

Ansell, C., & Gash, A. (2008). Collaborative governance in theory and practice. *Journal of Public Administration Research and Theory, 18*(4), 543–571.

Asheim, B. T., & Gertler, M. S. (2005). The geography of innovation: Regional innovation systems. In J. Fagerberg, D. C. Mowery, & R. R. Nelson (Eds.), *The Oxford handbook of innovation* (pp. 291–317). Oxford University Press.

Battilana, J., Leca, B., & Boxenbaum, E. (2009). How actors change institutions: Towards a theory of institutional entrepreneurship. *Academy of Management Annals, 3*(1), 65–107.

Becattini, G. (1990). The Marshallian industrial district as a socio-economic notion. In F. Pyke, G. Becattini, & W. Sengenberger (Eds.), *Industrial districts and inter-firm co-operation in Italy* (pp. 37–51). International Institute for Labour Studies.

Boschma, R. A. (2005). Proximity and innovation: A critical assessment. *Regional Studies, 39*(1), 61–74.

Boschma, R. A., & Iammarino, S. (2009). Related variety, trade linkages, and regional growth in Italy. *Economic Geography, 85*(3), 289–311.

Charnes, A., Cooper, W. W., & Rhodes, E. (1978). Measuring the efficiency of decision making units. *European Journal of Operational Research, 2*(6), 429–444.

Charron, N., Dijkstra, L., & Lapuente, V. (2014). Regional governance matters: Quality of government within European Union member states. *Regional Studies, 48*(1), 68–90.

Charron, N., Lapuente, V., & Bauhr, M. (2019). *Sub-national quality of government in EU member states: Presenting the 2017 European Quality of Government Index and its relationship with well-being.* European Commission, DG Regional and Urban Policy.

Cherchye, L., Moesen, W., Rogge, N., & Van Puyenbroeck, T. (2007). An introduction to 'benefit of the doubt' composite indicators. *Social Indicators Research, 82*(1), 111–145.

Cherchye, L., Moesen, W., Rogge, N., Van Puyenbroeck, T., Saisana, M., Saltelli, A., Liska, R., & Tarantola, S. (2008a). Creating composite indicators with DEA and robustness analysis: The case of the Technology Achievement Index. *Journal of the Operational Research Society, 59*(2), 239–251.

Cherchye, L., Ooghe, E., & Van Puyenbroeck, T. (2008b). Robust human development rankings. *The Journal of Economic Inequality, 6*(4), 287–321.

Coase, R. H. (1937). The nature of the firm. *Economica, 4*(16), 386–405.

Cohen, W. M., & Levinthal, D. A. (1990). Absorptive capacity: A new perspective on learning and innovation. *Administrative Science Quarterly, 35*(1), 128–152.

Coleman, J. S. (1990). *Foundations of social theory.* Harvard University Press.

Cooke, P. (1992). Regional innovation systems: Competitive regulation in the new Europe. *Geoforum, 23*(3), 365–382.

Corrente, S., Greco, S., Nicotra, M., Romano, M., & Schillaci, C. (2019). Evaluating and comparing entrepreneurial ecosystems using SMAA and SMAA-S. *The Journal of Technology Transfer, 44*(2), 485–519.

Cunningham, J., Del Giudice, M., Nicotra, M., O'Kane, C., & Romano, M. (2022). Principal investigators and knowledge management: A micro foundational conceptual framework. In D. Audretsch, E. Lehmann, & A. Link (Eds.), *Handbook of technology transfer* (pp. 57–74). Edward Elgar.

Del Giudice, M., Nicotra, M., Romano, M., & Schillaci, C. E. (2016). Entrepreneurial performance of principal investigators and country culture. *The Journal of Technology Transfer, 42*(2), 320–337.

DiMaggio, P. J. (1988). Interest and agency in institutional theory. In L. G. Zucker (Ed.), *Institutional patterns and organizations: Culture and environment* (pp. 3–21). Ballinger.

Dyer, J. H., & Singh, H. (1998). The relational view: Cooperative strategy and sources of interorganizational competitive advantage. *Academy of Management Review, 23*(4), 660–679.

Edwards, B., & McCarthy, J. D. (2004). Resources and social movement mobilization. In D. A. Snow, S. A. Soule, & H. Kriesi (Eds.), *The Blackwell companion to social movements* (pp. 116–152). Blackwell.

Emerson, K., Nabatchi, T., & Balogh, S. (2012). An integrative framework for collaborative governance. *Journal of Public Administration Research and Theory, 22*(1), 1–29.

Florida, R. (1995). Toward the learning region. *Futures, 27*(5), 527–536.

Frenken, K., Van Oort, F., & Verburg, T. (2007). Related variety, unrelated variety and regional economic growth. *Regional Studies, 41*(5), 685–697.

Fukuyama, F. (1995). *Trust: The social virtues and the creation of prosperity.* Free Press.

Gaaloul, H., & Khalfallah, S. (2014). Application of the 'benefit-of-the-doubt' approach for the construction of a digital access indicator: A revaluation of the 'Digital Access Index'. *Social Indicators Research, 118*(1), 45–56.

Garud, R., Hardy, C., & Maguire, S. (2007). Institutional entrepreneurship as embedded agency: An introduction to the special issue. *Organization Studies, 28*(7), 957–969.

Granovetter, M. (1985). Economic action and social structure: The problem of embeddedness. *American Journal of Sociology, 91*(3), 481–510.

Greco, S., Ishizaka, A., Tasiou, M., & Torrisi, G. (2019a). On the methodological framework of composite indices: A review of the issues of weighting, aggregation, and robustness. *Social Indicators Research, 141*(1), 61–94.

Greco, S., Ishizaka, A., Tasiou, M., & Torrisi, G. (2019b). Sigma-mu efficiency analysis: A methodology for evaluating units through composite indicators. *European Journal of Operational Research, 278*(3), 942–960.

Hall, P. A., & Soskice, D. (2001). *Varieties of capitalism: The institutional foundations of comparative advantage.* Oxford University Press.

Henriques, C., Sousa, S., & Viseu, C. (2024). Assessing the gender gap in the energy sector across European countries: Applying the benefit of the doubt approach. In N. S. Caetano (Ed.), *Sustainable development with renewable energy* (pp. 443–456). Springer.

Kale, P., Singh, H., & Perlmutter, H. (2000). Learning and protection of proprietary assets in strategic alliances: Building relational capital. *Strategic Management Journal, 21*(3), 217–237.

Kattel, R., & Mazzucato, M. (2018). Mission oriented innovation policy and dynamic capabilities in the public sector. *Industrial and Corporate Change, 27*(5), 787–801.

Lavie, D. (2006). The competitive advantage of interconnected firms: An extension of the resource-based view. *Academy of Management Review, 31*(3), 638–658.

Lundvall, B. A. (Ed.). (1992). *National systems of innovation: Toward a theory of innovation and interactive learning.* Pinter.

Lundvall, B. A., & Johnson, B. (1994). The learning economy. *Journal of Industry Studies, 1*(2), 23–42.

Marshall, A. (1890). *Principles of economics.* Macmillan.

Mazzucato, M. (2018). *The value of everything: Making and taking in the global economy.* Allen Lane.

McCarthy, J. D., & Zald, M. N. (1977). Resource mobilization and social movements: A partial theory. *American Journal of Sociology, 82*(6), 1212–1241.

Nelson, R. R. (Ed.). (1993). *National innovation systems: A comparative analysis.* Oxford University Press.

Nicotra, M., & Romano, M. (2018). *Ecosistemi della Conoscenza e absorptive capacity.* Giappichelli.

Nicotra, M., Romano, M., & Del Giudice, M. (2021). Fulfilling university third mission: Towards an ecosystemic strategy of entrepreneurship education. *Studies in Higher Education, 46*(5), 1000–1010.

Nicotra, M., Romano, M., Del Giudice, M., & Schillaci, C. E. (2018). The causal relation between entrepreneurial ecosystem and productive entrepreneurship. *The Journal of Technology Transfer, 43*(3), 640–673.

North, D. C. (1990). *Institutions, institutional change and economic performance*. Cambridge University Press.

Ostrom, E. (1990). *Governing the commons: The evolution of institutions for collective action*. Cambridge University Press.

Ostrom, E. (2010). Beyond markets and states: Polycentric governance of complex economic systems. *American Economic Review, 100*(3), 641–672.

Pedraja-Chaparro, F., Salinas-Jimenez, J., & Smith, P. (1997). On the role of weight restrictions in data envelopment analysis. *Journal of Productivity Analysis, 8*(2), 215–230.

Polanyi, K. (1944). *The great transformation: The political and economic origins of our time*. Farrar and Rinehart.

Porter, M. E. (1990). *The competitive advantage of nations*. Free Press.

Porter, M. E. (1998). Clusters and the new economics of competition. *Harvard Business Review, 76*(6), 77–90.

Provan, K. G., & Kenis, P. (2008). Modes of network governance: Structure, management, and effectiveness. *Journal of Public Administration Research and Theory, 18*(2), 229–252.

Putnam, R. D. (1993). *Making democracy work: Civic traditions in modern Italy*. Princeton University Press.

Putnam, R. D. (2000). *Bowling alone: The collapse and revival of American community*. Simon and Schuster.

Pyke, F., Becattini, G., & Sengenberger, W. (Eds.). (1992). *Industrial districts and inter-firm co-operation in Italy*. International Institute for Labour Studies.

Rodrik, D., Subramanian, A., & Trebbi, F. (2004). Institutions rule: The primacy of institutions over geography and integration in economic development. *Journal of Economic Growth, 9*(2), 131–165.

Rogge, N. (2011). Granting teachers the 'benefit of the doubt' in performance evaluations. *International Journal of Educational Management, 25*(6), 590–614.

Roll, Y., & Golany, B. (1993). Alternate methods of treating factor weights in DEA. *Omega, 21*(1), 99–109.

Romano, M., Catalfo, P., Nicotra, M. (2014). Science parks and intellectual capital: An integrated model for intangibles' representation, evaluation and control. *Journal Of Intellectual Capital*, 15, 537–553.

Schillaci, C. E., Romano, M., & Nicotra, M. (2013). Territory's absorptive capacity. *Entrepreneurship Research Journal, 3*(1), 109–126.

Schot, J., & Steinmueller, W. E. (2018). Three frames for innovation policy: R&D, systems of innovation and transformative change. *Research Policy, 47*(9), 1554–1567.

Shwartz, M., Burgess, J. F., & Berlowitz, D. (2009). Benefit-of-the-doubt approaches for calculating a composite measure of quality. *Health Services and Outcomes Research Methodology, 9*(4), 234–251.

Whitley, R. (1999). *Divergent capitalisms: The social structuring and change of business systems*. Oxford University Press.

Williamson, O. E. (1985). *The economic institutions of capitalism*. Free Press.

Woolcock, M. (1998). Social capital and economic development: Toward a theoretical synthesis and policy framework. *Theory and Society, 27*(2), 151–208.

Chapter 3
Transformative Purpose

A Transformative Purpose is the fundamental reason why a purpose ecosystem exists, the shared commitment that motivates its actors and gives direction to all its activities. The transition from enabling conditions to transformation also delineates a shift from what makes collective action possible to what shapes its meaning.

Transformative Purpose is not a strategy, a plan, or a set of governance rules. It is the social and environmental horizon toward which the ecosystem orients its collective energy.

The chapter is organized in three sections. Section 3.1 develops the environmental sustainability dimension of Transformative Purpose, tracing the evolution from firm-level environmental practices to the ecosystem-level transitions that represent the most ambitious expression of environmental commitment. Section 3.2 develops the social transformation dimension, examining what it means to genuinely pursue community building, equity, and shared well-being as organizational purposes. Section 3.3 introduces indicators of Transformative Purpose and so the output and short-term outcome to assess whether an ecosystem is moving in the right direction, and what distinguishes these proximate signals from the long-term impact measures developed in Chap. 6.

3.1 From Firm-Level Practices to Ecosystem-Level Environmental Sustainability

Environmental sustainability has changed profoundly as a concept over the past three decades. It began as a corporate compliance concern and it has become a strategic imperative. More and more scholars and practitioners see organizations strictly connected with the natural environment and the ecological systems. Environmental sustainability is considered related to every dimension of organizational life, from business models to investment decisions and stakeholder relationships

M. Nicotra and M. Romano, *Purpose Ecosystems*, Contributions to Management Science, https://doi.org/10.1007/978-3-032-38382-2_3

(Bansal & Roth, 2000; Dyllick & Hockerts, 2002; Schaltegger & Burritt, 2018). It is not a constraint on performance. It is a precondition for it.

In 1987, the Brundtland Commission defined sustainable development as "development that meets the needs of the present without compromising the ability of future generations to meet their own needs." Starting from this concept, management scholars have developed precise understandings better suited to guiding organizational action. Dyllick and Hockerts (2002) distinguish between ecological sustainability, that is, maintaining the natural capital stock that supports human and organizational activity, and the broader concept of sustainability that integrates economic, social, and environmental dimensions.

Just some years before, in 1997, Elkington had developed the triple bottom line framework, defining the principle that organizational performance must be assessed across all three dimensions simultaneously. In 2018, Elkington emphasized the importance of the three dimensions enabling genuine transformation rather than superficial reporting. Besides, Bansal & Roth (2000) identify the three motivations for corporate environmental responsiveness. These are competitiveness, legitimation, and ecological responsibility. However, they argue that only the third produces genuine, deep commitment rather than just compliance.

At the firm level, environmental sustainability relates to all the practices needed to reduce environmental footprint, improve resource efficiency, and contribute to ecological regeneration. The literature identifies several strategic orientations reflecting different levels of commitment. We can distinguish reactive organizations, that are those complying with environmental regulations and not going beyond compliance (Hart, 1995). Then proactive organizations, which develop environmental capabilities, generating competitive advantage through cost reduction, product differentiation, and reputational assets (Porter & van der Linde, 1995). Finally, systemic organizations that integrate environmental considerations into their business models. They treat ecological constraints not as externalities to be managed but as foundational principles that guide what and how they produce (Bocken et al., 2014; Geissdoerfer et al., 2017). In this context, a role is also assigned to digital technologies in enabling firms to develop and protect intellectual assets that support sustainable innovation (Orlando et al., 2020).

In purpose ecosystems, the commitment to environmental sustainability goes beyond what firm-level frameworks are able to capture. Such commitment is not about managing environmental risks, complying with regulations, or signaling legitimacy to stakeholders. It is about genuinely orienting the ecosystem's collective activities toward the preservation and regeneration of the ecological systems. At the level of the ecosystem, environmental sustainability should be seen as a coordinated collective action which requires a specific orientation, which we might call transformative. Organizations in an ecosystem manage their own environmental footprint but actively contribute to the transformation of the broader systems within which they operate, and so supply chains, industry norms, regulatory frameworks, and consumer behaviors.

Resource efficiency is one of the dimensions of firm-level environmental sustainability. Environmental impact can be reduced by the efficient use of energy, water, materials, and land. Not only, resource efficiency also generates cost reduction and productivity improvement (Ambec & Lanoie, 2008; Porter & van der Linde, 1995). Environmental management systems, most prominently ISO 14001, provide organizations with structured approaches to identifying, monitoring, and improving their environmental performance (Boiral & Henri, 2012). The analysis of resource efficiency of the individual firm can be extended, including the full environmental footprint of products and services across their entire lifecycle. This is known as the life cycle assessment (LCA) methodology (Guinee et al., 2011).

For purpose ecosystems, firm-level resource efficiency is a necessary but insufficient expression of environmental commitment. It reduces environmental damages but it does not generate the systemic change required by a Transformative Purpose.

Carbon management is a significant dimension of firm-level environmental sustainability given climate crisis. The Greenhouse Gas Protocol is a framework for corporate carbon accounting and it has been developed by the World Resources Institute and the World Business Council for Sustainable Development. It distinguishes Scope 1 emissions that are direct emissions from owned or controlled sources. Moreover Scope 2 emissions that are indirect emissions from purchased energy and Scope 3 emissions that are all other indirect emissions in the value chain (Ranganathan et al., 2004). The Science Based Targets initiative (SBTi) enables organizations to set emission reduction targets aligned with the Paris Agreement, limiting global warming to 1.5 °C.

For purpose ecosystems, carbon management encompasses the collective carbon footprint of the ecosystem as a whole, including the supply chain relationships, shared infrastructure, and territorial economic activities that ecosystem actors collectively influence.

The *circular economy* is another significant development in environmental sustainability management of the past decade. It is an alternative to the linear model typical of industrial production since the Industrial Revolution (Ellen MacArthur Foundation, 2013; Geissdoerfer et al., 2017). The circular economy aims to eliminate waste and to extend the life of products and materials. These principles have relevant implications for business model design, product architecture, supply chain organization, and the governance of material flows across industry boundaries.

For purpose ecosystems, the circular economy is particularly relevant because many of its most powerful applications involve the coordination of material flows across organizational boundaries. We refer to industrial symbiosis, where the waste of one organization becomes the resource input of another; product-as-a-service models, where manufacturers retain ownership of their products and recover them for remanufacturing at end of life; and collaborative consumption platforms, where shared use reduces the total volume of goods produced.

Biodiversity and ecosystem health have historically received less attention in the management literature than carbon and resource efficiency, but are rapidly gaining recognition as a strategic priority of equal importance. The Dasgupta (2021) on the economics of biodiversity explains the dependence of economic systems on natural capital. It argues that the degradation of biodiversity is a threat to long-term economic prosperity. The Taskforce on Nature-related Financial Disclosures (TNFD), launched in 2021, is developing a framework for organizations to assess and disclose their impacts on nature. For purpose ecosystems, biodiversity and ecosystem health are central dimensions of Transformative Purpose. Their commitment should go beyond the reduction of harm toward the active creation of ecological value. This is connected to rewilding, to sustainable land management, to nature-based solutions, and to the restoration of degraded ecosystems.

In purpose ecosystems, the transition from firm-level environmental practices to ecosystem-level transitions represents the most ambitious dimension of environmental sustainability. Ecosystem-level transitions are changes in the entire configuration of actors, technologies, institutions, and practices. No single organization can drive these transitions alone. They require multiple actors to work in a coordinated way toward a shared vision of change. They need also the right conditions to take hold and so institutional frameworks, financial mechanisms, and cultural norms that make the transition pathway not just desirable, but viable (Geels, 2002; Markard et al., 2012; Loorbach et al., 2017).

The *sustainability transitions literature*, which has developed extensively over the past two decades, explain how sociotechnical systems change in the direction of greater environmental sustainability (Geels, 2002; Rotmans et al., 2001; Loorbach, 2010). The multi-level perspective (MLP), developed by Geels (2002), distinguishes three levels: the landscape, that is, the macro-level context of long-term trends and deep cultural values. Moreover, the regime, that is, the established configuration of technologies, institutions, practices, and actor networks that constitutes the dominant system. And, finally, the niche, referring to the protected spaces where radical innovations emerge and develop before potentially challenging and transforming the regime. Transitions occur when niche innovations, amplified by landscape pressures such as climate change or resource scarcity, break through regime structures and reconfigure the system at the regime level. For purpose ecosystems, the most significant contribution to environmental sustainability lies not in the incremental improvement of existing practices but in the incubation, scaling, and diffusion of the niche innovations that can challenge and transform established sociotechnical regimes.

Strategic niche management (Kemp et al., 1998; Raven et al., 2010) provides practical guidance for how purpose ecosystems can create and manage the protected spaces that allow sustainability innovations to develop. Strategic niches are characterized by protective shielding from mainstream market pressures, learning processes that accelerate the development of novel technologies and practices, and the building of social networks connecting innovators with potential users, investors, and institutional supporters. Purpose ecosystems are particularly well-suited to this function because their multi-actor composition enables them to mobilize the diverse capabilities needed for niche development, their shared purpose provides the motivational foundation for sustained commitment to innovations that may take years to reach commercial viability, and their embeddedness in territorial communities gives them access to the local knowledge and social capital needed to adapt innovations to specific place-based contexts.

Transition management, developed by Loorbach and colleagues (2010) at the Dutch Research Institute for Transitions, provides a framework for navigating the complexity of sustainability transitions. Transition arenas, the multi-actor forums through which transition management is implemented in practice, share important structural features with purpose ecosystems. They bring together frontrunners from diverse institutional sectors, they operate with a shared long-term vision that transcends the interests of any single participant, and they pursue transformative change through visioning, experimentation, and strategic networking. The connection between transition management and purpose ecosystem theory is more than analogical. Purpose ecosystems can function as transition arenas, providing the collaborative infrastructure through which territorial actors develop and pursue shared environmental sustainability visions.

Planetary boundaries, developed by Rockstrom et al. (2009) and refined by Steffen et al. (2015), provide an important framework for understanding the ultimate limits within which environmental sustainability commitments must operate. The framework identifies nine Earth system processes. They include climate change, biodiversity loss, land-system change, freshwater use, and biogeochemical flows. Each of these represents a point beyond which the risk of irreversible disruption to the Earth system grows significantly. For purpose ecosystems, planetary boundaries provide a scientific grounding for environmental commitments and a hierarchy of priorities. The most urgent dimensions of environmental sustainability are those where human activities are already transgressing or approaching these limits.

The planetary boundaries concept has been extended by Raworth (2017) with the doughnut economics framework. The author combines planetary boundary with a social foundation and creates an integrated framework for sustainable development. The doughnut defines a safe and just space for humanity between a social foundation and an ecological ceiling. It merges the minimum standard of well-being and the planetary boundaries. This paradigm is particularly relevant for purpose ecosystems because it explicitly integrates environmental sustainability and social transformation as co-equal and interdependent dimensions of purpose. In this way, purpose ecosystems aim not merely to reduce environmental harm or expand social access, but to pursue a trajectory of development that respects ecological limits and advances social foundations. This is the integrated vision of Transformative Purpose.

Box 3.1 Key Frameworks for Environmental Sustainability in Purpose Ecosystems

Triple Bottom Line (Elkington, 1997). Organizations must be assessed across economic, social, and environmental performance simultaneously. Purpose ecosystems extend this logic to the ecosystem level, assessing collective performance across all three dimensions.

Natural Resource-Based View (Hart, 1995). Environmental capabilities, including pollution prevention, product stewardship, and sustainable development can sustain competitive advantage. Purpose ecosystems can develop collective environmental capabilities that exceed the sum of what individual organizations create.

Circular Economy (Ellen MacArthur Foundation, 2013). Economic systems keep in circulation products and materials through processes like maintenance, reuse, refurbishment, remanufacture, recycling, and composting. Purpose ecosystems can implement such principles across organizational boundaries.

Multi-Level Perspective on Sustainability Transitions (Geels, 2002). Transition is guided by the interaction of three interdependent levels, that are, landscape pressures, regime structures, and niche innovations. Purpose ecosystems can function as niches able to incubate and scale sustainability innovations.

Planetary Boundaries (Rockstrom et al., 2009). Biophysical limits must not be transgressed if organization want to avoid catastrophic Earth system changes. Purpose ecosystems must be oriented within these limits.

Doughnut Economics (Raworth, 2017). Sustainable development means also meeting social foundations for all people. In purpose ecosystems, this framework directly connects environmental sustainability and social transformation as co-equal dimensions of purpose.

The practical expression of environmental sustainability as Transformative Purpose takes different forms depending on the territorial context, the actor composition, and the specific environmental challenges the ecosystem addresses. In energy-intensive industrial territories, the transition to renewable energy and the decarbonization of industrial processes may be the primary focus. In agricultural territories, sustainable land management, biodiversity conservation, and the transition to regenerative farming practices may be central. In urban contexts, sustainable mobility, green infrastructure, circular waste management, and the decarbonization of buildings may define the environmental agenda. In coastal and marine territories, the sustainability of fisheries, the protection of marine ecosystems, and the management of coastal climate risks may be the defining concerns. What unites these diverse expressions of environmental commitment is the systemic orientation: the ambition to generate not merely incremental improvements in individual organizational practices but transformative change in the territorial systems within which those organizations operate.

Spotlight 3.1 The ECOSISTER Ecosystem in Emilia-Romagna

ECOSISTER (Ecosystem for Sustainable Transition in Emilia-Romagna) was launched in 2022 with funding from Italy's National Recovery and Resilience Plan, with the University of Bologna as lead institution and a budget of over 120 million euros distributed across six thematic spokes covering circular economy, energy, climate, environment, green manufacturing, and digital-ecological transition. On paper, it is an innovation ecosystem funded by public policy. In practice, it is one of the most compelling examples available of how policy intervention can create the structural conditions for Transformative

Purpose to emerge and become genuinely shared across a diverse multi-actor community.

The key to understanding ECOSISTER as a purpose ecosystem lies in its territorial context. Emilia-Romagna had already developed, before ECOSISTER was conceived, the Patto per il Lavoro e per il Clima. This is a territorial pact explicitly linking labor, economic development, and climate action, signed by the regional government, trade unions, employers' associations, universities, civic organizations, and local governments. This pact embedded ecological transition as a shared territorial commitment, not a government target. ECOSISTER did not generate this commitment. It found it already present and provided the resources and structures to realize it at scale.

The six spokes of ECOSISTER illustrate the two dimensions of Transformative Purpose. The environmental sustainability dimension is related to circular economy transitions that close material loops, energy transition activities that reduce fossil fuel dependence, and environmental monitoring that protects the ecological systems on which communities depend. But ECOSISTER also addresses the social transformation dimension, related to its commitment to combining digital transition and sustainability with work and people's well-being. The cascade funding calls directed specifically toward enterprises in Southern Italy (providing resources and connections to companies that lack access to the Emilia-Romagna innovation ecosystem) are a direct expression of the social transformation commitment embedded in its purpose.

The TTIP (Technology Transfer and Innovation Programme) that ECOSISTER has developed alongside its research activities is particularly significant. The TTIP explicitly includes startups, spin-offs, SMEs, civic organizations, public administrations, and citizens, not merely universities and large enterprises. The inclusion of citizens and civic organizations in innovation activities is rare in research ecosystems funded under Italian PNRR. In ECOSISTER, it reflects a genuine commitment to the principle that environmental sustainability transitions require community engagement, not merely technological development. What makes ECOSISTER instructive is precisely the tension at its heart where a policy-initiated ecosystem attempts to realize a Transformative Purpose whose full depth can only emerge from genuine multi-actor engagement. The evidence so far suggests this tension is being managed with more sophistication than is typical of policy-initiated innovation ecosystems.

3.2 Social Transformation: Building Communities, Equity, and Shared Well-Being

Social transformation is the second defining dimension of Transformative Purpose. Like environmental sustainability, it is a commitment and a direction. A shared orientation toward a more just, equitable, and flourishing social world that motivates ecosystem actors and gives meaning to their collective activities. And like environmental sustainability, it has a complex theoretical genealogy that spans multiple disciplines. It has been operationalized in many different ways, and carries important tensions that must be honestly confronted rather than dissolved through facile rhetoric.

The concept of social transformation in management scholarship draws on traditions from sociology, political economy, development studies, and social innovation that have not always been in productive dialogue with each other. First, we can refer to the sociology of social movements (McAdam et al., 1996; Tarrow, 2011). This is relevant to understand how collective action generates changes in social structures, norms, and power relations. Besides, the development economics and the capabilities approach (Nussbaum, 2011; Sen, 1999) provide frameworks for understanding genuine human development beyond income and consumption metrics. Also, the social innovation literature (Mulgan, 2019; Phills et al., 2008; Westley et al., 2011) sheds light on how new solutions to social challenges emerge, spread, and create systemic change. Finally, stakeholder theory (Freeman, 1984; Freeman et al., 2010) offers a theoretical lens to examine how organizations can create value for multiple constituencies simultaneously.

Drawing on all these traditions, this section develops an integrated understanding of social transformation as the purpose commitment to building communities, advancing equity, and generating shared well-being.

Community is the most fundamental unit of social transformation in purpose ecosystems. They should be seen not simply as aggregations of individuals. They are social systems with their own histories, cultures, institutions, relationships, and collective identities. Management scholarship on community has evolved significantly from early instrumentalist views, which treated communities primarily as markets or labor pools, toward richer understandings of communities as value-creating systems in their own right. The cluster theory developed by Porter (1998) highlighted the economic importance of geographic communities of firms and institutions, showing how proximity, shared knowledge, and relational trust generate advantages more than individual firms in isolation. Also, the work of Putnam (1993, 2000) on social capital demonstrated how trust, civic engagement, and networks of reciprocity shape not only individual well-being but collective economic performance and democratic governance. In this context, we can cite the social enterprise literature (Dees, 1998; Austin et al., 2006; Battilana & Lee, 2014; Schillaci et al., 2017). This has documented how organizations embedded in communities, drawing on local knowledge and relationships, can generate social value that market and state institutions fail to produce. Social entrepreneurs act as change catalysts within these ecosystems, mobilizing resources and relationships to challenge entrenched patterns of exclusion and disadvantage.

For purpose ecosystems, community is not merely the context in which they operate but a constitutive element of what they are. The co-creation logic that is central to purpose ecosystem identity requires that communities be genuine partners in the design, delivery, and governance of ecosystem activities rather than passive recipients of services designed by others. This commitment has profound implications for how social transformation is pursued. The definition of what social transformation means in a given territory cannot be imposed from outside. It must emerge from the community itself. The knowledge needed to design effective strategies is not primarily scientific or technical but is distributed across the community in the form of lived experience, local relationships, and contextual understanding. And the legitimacy of purpose ecosystem activities depends on whether communities recognize themselves in the ecosystem's purpose and feel genuinely empowered by its activities rather than merely served by them.

Equity is the second foundational concept of social transformation in purpose ecosystems. Equity is not the same as equality. Equality means treating everyone the same. Equity means ensuring that everyone has what they need to participate fully in social, economic, and civic life. This may require treating people differently in proportion to their different starting points, needs, and circumstances. The distinction matters enormously for purpose ecosystems, because their most challenging and most important work often involves addressing the deeply unequal distributions of opportunity, capability, and power that characterize the communities they serve. A rigorous framework for understanding equity as a dimension of social purpose is given by Sen (1999). His moral framework evaluates human welfare based on "capabilities" and defines equity in terms of the substantive freedoms that people enjoy. Sen (1999) argues that development should be understood as the expansion of substantive freedoms, that is, the real opportunities that people have to live the kinds of lives they have reason to value.

Nussbaum (2011) operationalizes the capabilities approach through a list of central human capabilities. In this way, she provides a framework for assessing whether social transformation initiatives are genuinely advancing equity or merely redistributing resources without changing the structural conditions that generate inequality.

Structural inequality is the primary target of genuine social transformation. Racial and ethnic inequality, gender inequality, disability-based exclusion, class-based opportunity gaps, and geographic marginalization are the result of individual deficiencies. They are produced by discriminatory housing markets, unequal educational systems, labor market practices that systematically disadvantage certain groups, cultural norms that devalue certain identities, and political institutions that amplify the voices of some, silencing others. Purpose ecosystems committed to genuine social transformation must therefore engage with structural inequality directly, not merely providing services to the individuals who suffer its consequences but working to change the structures that generate those consequences. This is what distinguishes transformative social purpose from philanthropic service delivery. The aspiration is not merely to help individuals navigate unjust systems but to change the systems themselves.

Shared well-being extends the focus of social transformation beyond the reduction of deprivation and the advancement of equity toward the positive conditions that enable human communities to flourish. The well-being literature in economics and psychology has documented that income and consumption (the primary metrics of conventional economic development) are poor predictors of subjective well-being and human flourishing (Diener et al., 2010; Layard, 2005; Stiglitz et al., 2009). The Better Life Framework of OECD (2011) identifies eleven dimensions of well-being that go beyond income. These include housing, work, health, education, civic engagement, social connections, safety, life satisfaction, environment quality, work-life balance, and community. Layard (2005), in his happiness economics, demonstrates that the stronger predictors of individual well-being are mental health, social relationships, and sense of purpose. He suggests that strategies focused exclusively on economic empowerment and that neglect relational and psychological dimensions will underestimate what human flourishing requires.

For purpose ecosystems, shared well-being emphasizes the collective and relational dimensions of human flourishing. People do not flourish in isolation. Their well-being is fundamentally dependent on the quality of their relationships, the vitality of the communities they belong to, and the health of the social fabric that sustains those communities. Social capital (Coleman, 1990; Putnam, 1993) is a fundamental dimension of shared well-being. Communities with high social capital are more resilient in the face of adversity, more effective at coordinating collective responses to shared challenges, and generate greater well-being for their members than communities with low social capital. Cultural vitality (the capacity of communities to maintain, develop, and transmit their cultural identities, practices, and forms of expression) is a dimension of shared well-being frequently overlooked in economic development frameworks but central to the sense of belonging, identity, and meaning that sustains human communities over time. For purpose ecosystems operating in territories with diverse cultural identities (including indigenous communities, minority language groups, and migrant communities) cultural vitality is not a peripheral concern but a fundamental dimension of the social transformation they are committed to generating.

The social innovation literature provides an important complement to the frameworks described above, offering a process-oriented understanding of how social transformation is generated in practice. Social innovation is defined as 'a novel solution to a social problem, that is, more effective, efficient, sustainable, or just than present solutions and for which the value created accrues primarily to society as a whole rather than private individuals' (Phills et al., 2008). In 2019, a comprehensive framework for social innovation was developed by Mulgan. In the framework, the author includes new ideas, organizational forms, and ways of working that meet social needs more effectively than existing approaches.

The conditions under which social innovations achieve systemic impact are analyzed by Westley et al. (2011). They argue that transformative change requires not only the development of effective new solutions but the deliberate strategy of scaling those solutions in ways that challenge and transform the systems that produce the problems they address.

The social innovation literature has, however, also attracted significant critical scrutiny in recent years. A first critique concerns what recent scholarship calls the "social innovation trap" (George et al., 2023). It is the tendency of the field to be fragmented into disciplinary silos in ways that obscure the structural conditions generating the social problems. Another critique concerns the relationship between social innovation and power. Moulaert et al. (2013) and Fraser (2013) argue that social innovation initiatives risk reinforcing rather than challenging existing power structures. More recently, Pel et al. (2020) have proposed the concept of transformative social innovation specifically to distinguish initiatives that generate genuine systemic change from those that produce incremental improvements without challenging the structural social problems.

Besides, Kania and Kramer (2011) provide a framework for collective impact. They explain how multi-actor collaborations can generate social transformation at scale.

Collective impact initiatives bring together organizations from different sectors around a common agenda, shared measurement systems, mutually reinforcing activities, continuous communication, and backbone support organizations. The framework has been influential in social innovation practice and has generated important insights about the governance requirements for multi-actor social transformation efforts. For purpose ecosystems, the collective impact framework offers useful methodological guidance, but the purpose ecosystem concept goes beyond it in several important respects. It emphasizes the emergent rather than designed character of shared purpose. Moreover, it insists on the co-creation role of communities rather than treating them as beneficiaries of institutional collaboration, and it aspires to systemic transformation rather than the improvement of outcomes within existing systems.

Community development theory and practice provides a third body of knowledge directly relevant to social transformation in purpose ecosystems. The community development tradition (Alinsky, 1971, Freire, 1970, Kretzmann & McKnight, 1993) emphasizes the importance of building on community assets rather than focusing exclusively on deficits, developing community leadership and self-determination, and working toward structural change in the conditions that constrain community development.

The asset-based community development (ABCD) approach, developed by Kretzmann & McKnight (1993), is a methodology for identifying and mobilizing the assets of communities. They refer to the skills, knowledge, relationships, and organizational resources that communities already possess and that can serve as the foundation for locally driven development. For purpose ecosystems, ABCD can be particularly valuable because it aligns with the co-creation logic. A purpose ecosystem that practices ABCD can engage communities in identifying their own assets, articulating their own visions, and designing their own pathways to the social transformation they seek.

Within this line of inquiry, we can also include the Participatory Action Research (PAR), developed across multiple disciplinary traditions from Lewin (1946) to Fals Borda (1987) to Reason and Bradbury (2008). It provides a methodology for generating new knowledge and change through the active involvement of community members as co-researchers. PAR insists that the most useful knowledge for social transformation is produced through a combination of theoretical frameworks developed by researchers and practical knowledge and experience of communities. For purpose ecosystems, PAR offers a model where learning and adaptation strengthen the community from within.

Coherently, we can also cite the tradition of the feminist political economy. It is associated with scholars such as Fraser (1995, 2013); Nussbaum (2011), and Young (1990), who provide crucial insights about the dimensions of social transformation. Fraser (1995) focuses on redistribution, recognition, and representation as necessary for genuine social transformation. Redistribution is related to changing the economic structures that generate inequality, recognition to changing the cultural norms that devalue certain identities and groups, representation to changing the political structures that exclude certain voices from decision-making. Redistribution without recognition reproduces the stigmatization of disadvantaged groups even if it improves their economic conditions. Recognition without redistribution fails to address the material foundations of inequality. Redistribution and recognition without representation prevent communities from defining and pursuing their own versions of social transformation. Young (1990) provides a comprehensive taxonomy of the forms of injustice that social transformation must address. The author lists the five faces of oppression: exploitation, marginalization, powerlessness, cultural imperialism, and violence. Similarly, Nussbaum (2011) provides a list of what social transformation ultimately aims to do. It encompasses life and so living to a normal length and avoiding premature death. In addition, bodily health and integrity, that means good health, adequate nourishment, and shelter and, at the same time freedom of movement and security from violence. The list also includes senses, imagination and thought to experience and create, emotions to experience attachments and love, practical reason for a conception of the good and for critical reflection. Other elements are affiliation and so living with and toward others, but also living with concern for and in relation to animals and the natural world. At the same time laughing, playing, and enjoying recreational activities as well as controlling over environment, that is, political participation and economic opportunity.

For purpose ecosystems, this list serves as a comprehensive evaluative framework. Genuine social transformation must advance all these capabilities, not only the economic ones

Box 3.2 Key Frameworks for Social Transformation in Purpose Ecosystems
Capabilities approach (Nussbaum, 2011; Sen, 1999). Development means the expansion of substantive freedoms and capabilities, not merely income growth. Purpose

ecosystems aim to expand the full range of human capabilities in the communities they serve.

Social capital (Coleman, 1990; Putnam, 1993). Networks, norms, and trust enable collective action and generate community well-being. Purpose ecosystems depend on and contribute to the social capital of their territories.

Three-dimensional justice (Fraser, 1995). Genuine equity requires redistribution of economic resources, recognition of cultural identities, and representation in political processes. Purpose ecosystems must address all three dimensions.

Asset-based community development (Kretzmann & McKnight, 1993). Communities possess assets, not only deficits. Social transformation builds on existing community strengths rather than imposing external solutions to externally defined problems.

Collective impact (Kania & Kramer, 2011). Large-scale social change requires multi-actor collaboration around a common agenda, shared measurement, mutually reinforcing activities, and backbone support. Purpose ecosystems share structural features with collective impact initiatives.

Systemic change (Mulgan, 2019; Westley et al., 2011). Genuine social innovation requires not only developing effective new solutions but scaling them in ways that challenge and transform the systems that produce social problems.

The practical expression of social transformation as Transformative Purpose takes different forms depending on the specific challenges, assets, and aspirations of the communities involved. As Sen (1999) underlined, human needs and aspirations differ profoundly from one another; no universal solutions exist that can be applied to all. This is true also for territories and communities. In territories that are economically disadvantaged, the primary focus can be creating quality employment, supporting entrepreneurship, developing skills and capabilities, and building economic assets. In territories with low cultural identities, social transformation may center on cultural vitality. In territories marked by unequal access to services, social transformation may focus on accessibility and inclusion, such as reducing barriers to healthcare, education, housing, and civic participation. In territories with high levels of social fragmentation and low social capital, the priority may be simply rebuilding associational life, relationship by relationship. What unites these diverse expressions is the shared orientation toward equity, community agency, and the conditions for genuine human flourishing.

Spotlight 3.2 The Transition Towns Network

In 2006, Rob Hopkins, a permaculture teacher in the small English town of Totnes in Devon, organized a community event to explore what the town might look like if it had reduced its energy consumption by half and developed its own local food system. The event was not the result of a policy mandate, a funding call, or an institutional initiative. It was the response of a community that had begun to feel, viscerally and collectively, the anxiety of fossil fuel dependence and climate vulnerability. What began as a conversation became a community project, the Transition Town Totnes, which became a model, which became a movement. By 2016, Transition Network had registered initiatives in over 50 countries, connecting more than 1000 communities across six continents.

The Transition Towns movement is perhaps the purest available example of a purpose ecosystem whose Transformative Purpose is genuinely inseparable from environmental sustainability and social transformation. The environmental dimension is explicit in the transition from fossil fuel dependence to local resilience. But the social transformation dimension is equally constitutive. Transition initiatives do not treat environmental change as a technical problem to be solved by engineers and policymakers. They treat it as a cultural and relational challenge that requires communities to rediscover their capacity for collective action, mutual care, and place-based identity.

The organizational form of the Transition movement illustrates the anatomy of an emergent purpose ecosystem with remarkable clarity. There is no headquarters that directs local initiatives. There is a Transition Network that provides resources, connections, training, and a shared framework, but it explicitly positions itself as a support structure, not a command center. Each local initiative develops its own projects, governance arrangements, and priorities in response to its specific community context. In Totnes, the initiative has developed a local currency, a community energy cooperative, urban food growing projects, and an inner transition program addressing the psychological and relational dimensions of cultural change. In other communities, the emphasis falls differently, but the shared commitment to ecological resilience and community agency remains the common thread.

Community members who participate in Transition initiatives report changes not only in their environmental practices but in their relationships with neighbors, their sense of agency and efficacy, and their understanding of what makes a good community. These relational and cultural changes are among the most durable contributions that purpose ecosystem can make to the communities they serve, and they are largely invisible to conventional impact measurement frameworks. The challenge the movement faces (maintaining authentic local ownership and community rootedness and, at the same time, developing the shared learning and collective accountability that enable systemic impact) is one that every purpose ecosystem with transformative ambitions must ultimately confront.

3.3 Outputs and Short-Term Outcomes Indicators of Transformative Purpose

Purpose ecosystems emerge from the bottom, from the interaction of actors who share values and a vision of social and environmental change. Their shared purpose emerges gradually through collective action, crystallizes as stakeholders discover common ground, and continues to evolve as the ecosystem learns and adapts. This emergent character has an important implication for indicators. In a conventional organization with a formal structure and a designated leadership, indicators are produced by the organization itself, reported to its governance bodies, and used to guide strategic decisions. In a purpose ecosystem that emerges organically from a territory, there is no single actor with the authority and the capacity to produce comprehensive indicators of ecosystem performance (Table 3.1).

Who, then, observes and interprets the signals that tell us whether a purpose ecosystem is moving in the right direction? And what are those signals?

Rather than designing a measurement system that an organization implements, we try to identify the observable signals that emerge naturally from the activities of a functioning purpose ecosystem and that can be detected by a range of observers. Researchers studying the ecosystem, policymakers evaluating their investments, community members can assess whether the changes they sought are occurring. In this section, we deal with outputs and short-term outcomes, the immediate, observable products of ecosystem activities and the first signals of change in conditions, behaviors, and capabilities that those activities generate. Outputs and short-term outcomes tell us that something is happening. Differently, impact tells us whether the trajectory of change is genuinely transformative. Chapter 6 will focus on impact, the long-term, systemic, and often difficult-to-attribute changes in social and environmental conditions that represent the ultimate test of Transformative Purpose.

In ecosystems that emerge organically from territorial actor networks, indicators are primarily observational. They are signals that external observers can detect to assess whether the ecosystem is developing in ways consistent with its Transformative Purpose. They are not produced by a central governance structure because no such structure exists. They are distributed across the ecosystem in the form of individual organizational reports, community surveys, environmental monitoring data, and the narratives that ecosystem actors share about their experiences.

In ecosystems that are initiated and supported by policy instruments (such as the PNRR-funded national ecosystems in Italy, the European Extended Partnerships, and the territorial development programs associated with Smart Specialization Strategies), the situation is different. These ecosystems have formal governance structures, contractual accountability relationships, and reporting obligations that make the systematic production of indicators possible and required. For policy-initiated ecosystems, the indicator framework can serve as the basis for formal monitoring and evaluation systems.

Outputs are the immediate, directly attributable products of ecosystem activities: the programs it delivers, the partnerships it forms, the innovations it develops, the

Table 3.1 Output and short-term outcome indicators

Output	Short-term outcome	Observation method
Environmental sustainability		
Number of organizations adopting environmental management systems (ISO 14001, EMAS)	Reduction in aggregate energy consumption across ecosystem organizations	Administrative records; energy bills
Volume of renewable energy generated or procured by ecosystem actors	Share of ecosystem energy consumption from renewable sources	Energy supplier data; organizational reports
Number of circular economy initiatives implemented across ecosystem	Reduction in waste generated per unit of output across ecosystem organizations	Waste management records; LCA data
Area of land under sustainable management or ecological restoration	Measurable improvement in biodiversity indicators in target areas	Field surveys; satellite monitoring
Number of carbon reduction plans developed and adopted	Scopes 1 and 2 emission reductions across participating organizations	GHG protocol reporting; third-party verification
Number of nature-based solutions designed and implemented	Ecosystem services delivered (carbon sequestration, water regulation, habitat provision)	Ecosystem services assessment tools
Number of environmental knowledge products created and disseminated	Uptake of environmental knowledge by organizations and policy bodies	Downloads; citations; policy references
Social transformation		
Number of community co-design processes conducted	Increase in community agency scores among participants	Participatory surveys; community self-assessment
Number of capability-building programs delivered	Measurable capability improvements among program participants	Pre/post assessments; follow-up surveys
Number of community organizations supported or created	Increase in associational density in target communities	Community mapping; administrative registers
Number of participatory governance mechanisms established	Increase in community representation in decision-making bodies	Governance records; stakeholder surveys
Number of cultural vitality initiatives implemented	Increase in cultural participation and cultural confidence among community members	Cultural participation surveys; program attendance
Number of quality jobs created in target communities	Employment rate improvements; wage adequacy among ecosystem-connected workers	HR records; labor market data

(continued)

Table 3.1 (continued)

Output	Short-term outcome	Observation method
Number of services made accessible to previously excluded groups	Reduction in identified accessibility barriers; increase in service uptake by target groups	Accessibility audits; service utilization data
Number of social enterprises and community organizations accessing ecosystem resources	Organizational sustainability of supported enterprises; mission delivery rates	Organizational surveys; financial accounts
Territorial development		
Number of territorial knowledge products created (reports, mappings, diagnoses)	Use of territorial knowledge in policy and planning decisions	Policy document analysis; stakeholder interviews
Number of inter-sector collaborations initiated in the territory	Depth and durability of new collaborative relationships	Network analysis; collaboration surveys
Number of territorial innovation initiatives supported	Adoption of innovations by organizations in the territory	Diffusion tracking; organizational surveys
Investment mobilized for territorial sustainability transitions	New investment flows toward sustainability-aligned activities	Financial data; investment registers
Number of policy dialogues and advocacy initiatives conducted	Changes in territorial policy frameworks aligned with transformative purpose	Policy monitoring; stakeholder assessment
Ecosystem vitality		
Number of new actors joining the ecosystem network	Diversity of actor types represented in the ecosystem	Network mapping; participation records
Number of collaborative projects and joint initiatives	Quality and depth of inter-organizational relationships	Collaboration surveys; project evaluations
Number of shared learning events, exchanges, and knowledge products	Evidence of learning integration in organizational practices	Learning assessments; practice change surveys
Number of community members actively engaged in ecosystem activities	Increase in community ownership and co-creation of ecosystem activities	Participation tracking; community surveys
Volume of resources mobilized and shared within the ecosystem	Efficiency gains from resource-sharing; reduction in duplication	Resource mapping; cost-benefit analysis

Source Authors' own elaboration

services it provides, the events it organizes, the knowledge it creates and disseminates. Outputs are observable in the short term, directly causally connected to ecosystem activities, and relatively easy to count and verify. They tell us that the ecosystem is active and producing results, but they do not tell us whether those results are generating meaningful change in the conditions that Transformative Purpose is oriented toward.

Short-term outcomes are the changes in conditions, behaviors, capabilities, and relationships that result from ecosystem outputs over a period of months to a few years. They represent the first evidence that ecosystem activities are working as intended and so programs are changing behaviors, partnerships are generating collaborative learning, services are improving capabilities, innovations are being adopted. Short-term outcomes are less directly attributable to ecosystem activities than outputs because other factors also influence the conditions being measured. They require more sophisticated data collection approaches, including survey methods, behavioral observation, and administrative data analysis. They are more informative about ecosystem effectiveness than outputs. They tell us not just that the ecosystem is active but that its activities are beginning to generate the changes in conditions that Transformative Purpose requires.

The framework presented below organizes outputs and short-term outcomes across four domains that are environmental sustainability, social transformation, territorial development, and ecosystem vitality. The first three domains correspond directly to the substantive content of Transformative Purpose. The fourth, ecosystem vitality, captures the relational and organizational health of the ecosystem itself. It concerns the growth of collaborative relationships, the development of shared learning, and the expansion of the actor network. These ecosystem vitality indicators matter because a purpose ecosystem, that is, achieving environmental and social results in the short term but whose collaborative fabric is fraying is unlikely to sustain its trajectory toward long-term transformative impact.

Several methodological principles should guide the use of this indicator framework, whether by external observers assessing an organically emergent ecosystem or by governance structures monitoring a policy-initiated one.

The tables above present a comprehensive reference architecture, not a minimum requirement. No ecosystem should attempt to monitor all these indicators simultaneously. The appropriate indicator set depends on the ecosystem's specific Transformative Purpose, the resources available for monitoring, and the questions that its actors, funders, or evaluators most need to answer. A good monitoring system typically includes five to ten carefully selected indicators that cover the most important dimensions of Transformative Purpose in that specific context, supplemented by qualitative evidence from community engagement processes.

Moreover, the most meaningful indicators of Transformative Purpose are those that communities themselves find informative and legitimate. The design of indicator frameworks should involve community members as active participants, not merely as data sources. This is particularly important for social transformation indicators, where the relevant dimensions of change are often best understood through the lived experience of community members rather than through external administrative data.

Participatory monitoring approaches are those in which community members define what success looks like, collect data on their own progress, and interpret the results in light of their own knowledge of local conditions. Such approaches generate indicators that are more relevant, more trusted, and more likely to influence ecosystem activities than externally designed measurement systems.

In addition, outputs and short-term outcomes are navigational signals, not verdicts. They tell us whether the ecosystem appears to be moving in the right direction, whether its activities are generating the expected early results, and whether there are signals of concern that warrant closer attention. Ecosystems that focus exclusively on optimizing their output and short-term outcome metrics at the expense of longer-term systemic changes are engaging in exactly the kind of indicator fixation that the broader measurement literature has warned against (Espeland & Sauder, 2007; Power, 1999). Indicators should be used to navigate, not to substitute for substantive reflection on whether Transformative Purpose is genuinely being advanced.

The same indicator can mean very different things in different territorial contexts. An increase in the number of community organizations in one territory may signal a genuine flourishing of associational life; in another, it may reflect the proliferation of organizationally fragile groups that will not survive without external support. A reduction in waste generation in one industrial context may represent genuine circular economy progress; in another, it may reflect production cuts driven by economic decline rather than sustainability strategy. Indicators should always be interpreted in light of a rich contextual understanding of the territory, the ecosystem, and the specific social and environmental challenges being addressed. Quantitative indicators should always be accompanied by qualitative evidence from community engagement and narrative documentation.

When a purpose ecosystem is initiated, structured, and funded by a policy instrument, the indicator framework presented above can and should be operationalized as a formal monitoring and evaluation system. In this context, indicators become accountability tools as well as navigational signals. They demonstrate to funders and policymakers that public investments are generating the expected results and contribute to the evidence base for future policy design. Policy-initiated ecosystems should invest in monitoring systems that are rigorous, transparent, and accessible to all stakeholders. The monitoring data generated by these systems can also serve a learning function, helping ecosystem actors to understand which activities are generating the strongest results and to redirect resources accordingly.

Box 3.3 Outputs, Short-Term Outcomes, and Impact

Outputs: The immediate, directly attributable products of ecosystem activities. Produced in the short term. Directly causally connected to ecosystem actions. Examples are number of programs delivered, number of partnerships formed, hectares under sustainable management, tons of CO_2 reduced.

Short-Term Outcomes: The first changes in conditions, behaviors, and capabilities that result from ecosystem outputs. Observable within months to a few years. Less directly attributable but more informative than outputs. Examples are improvement in community agency scores, reduction in energy consumption, increase in biodiversity indicators, growth of associational density.

Impact (Chap. 6): The long-term, systemic, and often emergent changes in social and environmental conditions that represent the ultimate test of Transformative Purpose. Observable over years to decades. Require sophisticated attribution methodologies. Examples are permanent reduction in territorial inequality, transition to a low-carbon energy system, restoration of ecosystem health, transformation of institutional practices toward equity.

In organically emergent ecosystems, outputs and short-term outcomes are observational signals rather than formally produced metrics. In policy-initiated ecosystems, they can and should be systematically monitored as part of formal accountability and learning systems. They should be used for navigation, not for judgment about the ultimate achievement of Transformative Purpose.

References

Alinsky, S. D. (1971). *Rules for radicals: A pragmatic primer for realistic radicals.* Random House.

Ambec, S., & Lanoie, P. (2008). Does it pay to be green? A systematic overview. *Academy of Management Perspectives, 22*(4), 45–62.

Austin, J., Stevenson, H., & Wei-Skillern, J. (2006). Social and commercial entrepreneurship: Same, different, or both? *Entrepreneurship Theory and Practice, 30*(1), 1–22.

Bansal, P., & Roth, K. (2000). Why companies go green: A model of ecological responsiveness. *Academy of Management Journal, 43*(4), 717–736.

Battilana, J., & Lee, M. (2014). Advancing research on hybrid organizing: Insights from the study of social enterprises. *Academy of Management Annals, 8*(1), 397–441.

Bocken, N. M. P., Short, S. W., Rana, P., & Evans, S. (2014). A literature and practice review to develop sustainable business model archetypes. *Journal of Cleaner Production, 65*, 42–56.

Boiral, O., & Henri, J. F. (2012). Modelling the impact of ISO 14001 on environmental performance: A comparative approach. *Journal of Environmental Management, 99*, 84–97.

Brundtland Commission. (1987). *Our common future.* Oxford University Press.

Coleman, J. S. (1990). *Foundations of social theory.* Harvard University Press.

Dasgupta, P. (2021). *The economics of biodiversity: The Dasgupta review.* HM Treasury.

Dees, J. G. (1998). *The meaning of social entrepreneurship.* Kauffman Foundation.

Diener, E., Lucas, R., Schimmack, U., & Helliwell, J. (2010). *Well-being for public policy.* Oxford University Press.

Dyllick, T., & Hockerts, K. (2002). Beyond the business case for corporate sustainability. *Business Strategy and the Environment, 11*(2), 130–141.

Elkington, J. (1997). *Cannibals with forks: The triple bottom line of twenty-first century business.* Capstone.

Elkington, J. (2018). 25 years ago I coined the phrase "triple bottom line." Here's why it's time to rethink it..

Ellen MacArthur Foundation. (2013). *Towards the circular economy: Economic and business rationale for an accelerated transition.* Ellen MacArthur Foundation.

Espeland, W. N., & Sauder, M. (2007). Rankings and reactivity: How public measures recreate social worlds. *American Journal of Sociology, 113*(1), 1–40.

Fals Borda, O. (1987). The application of participatory action research in Latin America. *International Sociology, 2*(4), 329–347.

Fraser, N. (1995). From redistribution to recognition? Dilemmas of justice in a post-socialist age. *New Left Review, 212*, 68–93.

Fraser, N. (2013). *Fortunes of feminism: From state-managed capitalism to neoliberal crisis*. Verso.

Freeman, R. E. (1984). *Strategic management: A stakeholder approach*. Pitman.

Freeman, R. E., Harrison, J. S., Wicks, A. C., Parmar, B. L., & de Colle, S. (2010). *Stakeholder theory: The state of the art*. Cambridge University Press.

Freire, P. (1970). *Pedagogy of the oppressed*. Herder and Herder.

Geels, F. W. (2002). Technological transitions as evolutionary reconfiguration processes: A multi-level perspective and a case-study. *Research Policy, 31*(8–9), 1257–1274.

Geissdoerfer, M., Savaget, P., Bocken, N. M. P., & Hultink, E. J. (2017). The circular economy: A new sustainability paradigm? *Journal of Cleaner Production, 143*, 757–768.

George, G., Haas, M. R., McGahan, A. M., Schillebeeckx, S. J. D., & Tracey, P. (2023). Purpose in the for-profit firm: A review and framework for management research. *Journal of Management, 49*(6), 1841–1869.

Guinee, J. B., Heijungs, R., Huppes, G., Zamagni, A., Masoni, P., Buonamici, R., Ekval, T., & Rydberg, T. (2011). Life cycle assessment: Past, present, and future. *Environmental Science and Technology, 45*(1), 90–96.

Hart, S. L. (1995). A natural-resource-based view of the firm. *Academy of Management Review, 20*(4), 986–1014.

Kania, J., & Kramer, M. (2011). Collective impact. *Stanford Social Innovation Review, 9*(1), 36–41.

Kemp, R., Schot, J., & Hoogma, R. (1998). Regime shifts to sustainability through processes of niche formation: The approach of strategic niche management. *Technology Analysis and Strategic Management, 10*(2), 175–198.

Kretzmann, J. P., & McKnight, J. L. (1993). Building communities from the inside out: A path toward finding and mobilizing a community's assets. *ACTA*. Publications.

Layard, R. (2005). *Happiness: Lessons from a new science*. Penguin.

Lewin, K. (1946). Action research and minority problems. *Journal of Social Issues, 2*(4), 34–46.

Loorbach, D. (2010). Transition management for sustainable development: A prescriptive, complexity-based governance framework. *Governance, 23*(1), 161–183.

Loorbach, D., Frantzeskaki, N., & Avelino, F. (2017). Sustainability transitions research: Transforming science and practice for societal change. *Annual Review of Environment and Resources, 42*, 599–626.

Markard, J., Raven, R., & Truffer, B. (2012). Sustainability transitions: An emerging field of research and its prospects. *Research Policy, 41*(6), 955–967.

McAdam, D., McCarthy, J. D., & Zald, M. N. (1996). *Comparative perspectives on social movements*. Cambridge University Press.

Moulaert, F., MacCallum, D., Mehmood, A., & Hamdouch, A. (2013). *The international handbook on social innovation: Collective action, social learning and transdisciplinary research*. Edward Elgar Publishing.

Mulgan, G. (2019). *Social innovation: How societies find the power to change*. Policy Press.

Nussbaum, M. C. (2011). *Creating capabilities: The human development approach*. Harvard University Press.

OECD. (2011). *How's life? Measuring well-being*. OECD Publishing.

Orlando, B., Mazzucchelli, A., Usai, A., Nicotra, M., & Paoletti, F. (2020). Are digital technologies killing future innovation? The curvilinear relationship between digital technologies and firm's intellectual property. *Journal of Intellectual Capital, 22*(3), 587–609.

Pel, B., Wittmayer, J. M., Dorland, J., & Søgaard Jørgensen, M. (2020). Unpacking the social innovation ecosystem: An empirically grounded typology of empowering network constellations. *Innovation: The European Journal of Social Science Research, 33*(3), 1–26.

Phills, J. A., Deiglmeier, K., & Miller, D. T. (2008). Rediscovering social innovation. *Stanford Social Innovation Review, 6*(4), 34–43.

Porter, M. E. (1998). Clusters and the new economics of competition. *Harvard Business Review, 76*(6), 77–90.

Porter, M. E., & van der Linde, C. (1995). Toward a new conception of the environment-competitiveness relationship. *Journal of Economic Perspectives, 9*(4), 97–118.

Power, M. (1999). *The audit society: Rituals of verification*. Oxford University Press.

Putnam, R. D. (1993). *Making democracy work: Civic traditions in modern Italy*. Princeton University Press.

Putnam, R. D. (2000). *Bowling alone: The collapse and revival of American community*. Simon and Schuster.

Ranganathan, J., Corbier, L., Bhatia, P., Schmitz, S., Gage, P., & Oren, K. (2004). *The greenhouse gas protocol: A corporate accounting and reporting standard*. World Resources Institute and World Business Council for Sustainable Development.

Raven, R., Van den Bosch, S., & Weterings, R. (2010). Transitions and strategic niche management: Towards a competence kit for practitioners. *International Journal of Technology Management, 51*(1), 57–74.

Raworth, K. (2017). Doughnut economics: Seven ways to think like a twenty-first-century economist..

Reason, P., & Bradbury, H. (Eds.). (2008). *The SAGE handbook of action research: Participative inquiry and practice*. SAGE.

Rockstrom, J., Steffen, W., Noone, K., Persson, A., Chapin, F. S., Lambin, E. F., Lenton, T. M., Scheffer, M., Folke, C., Schellnhuber, H. J., Nykvist, B., de Wit, C. A., Hughes, T., van der Leeuw, S., Rodhe, H., Sorlin, S., Snyder, P. K., Costanza, R., Svedin, U., Falkenmark, M., Karlberg, L., Corell, R. W., Fabry, V. J., Hansen, J., Walker, B., Liverman, D., Richardson, K., Crutzen, P., & Foley, J. A. (2009). A safe operating space for humanity. *Nature, 461*(7263), 472–475.

Rotmans, J., Kemp, R., & van Asselt, M. (2001). More evolution than revolution: Transition management in public policy. *Foresight, 3*(1), 15–31.

Schaltegger, S., & Burritt, R. (2018). Business cases and corporate engagement with sustainability: Differentiating ethical motivations. *Journal of Business Ethics, 147*(2), 241–259.

Schillaci, C. E., Romano, M., & Nicotra, M. (2017). Social entrepreneurs as change catalysts: Evidences from case studies. *International Journal of Managerial and Financial Accounting, 9*(2), 124–139.

Sen, A. (1999). *Development as freedom*. Oxford University Press.

Steffen, W., Richardson, K., Rockstrom, J., Cornell, S. E., Fetzer, I., Bennett, E. M., Biggs, R., Carpenter, S. R., de Vries, W., de Wit, C. A., Folke, C., Gerten, D., Heinke, J., Mace, G. M., Persson, L. M., Ramanathan, V., Reyers, B., & Sorlin, S. (2015). Planetary boundaries: Guiding human development on a changing planet. *Science, 347*(6223).

Stiglitz, J. E., Sen, A., & Fitoussi, J. P. (2009). *Report by the commission on the measurement of economic performance and social progress*. Commission on the Measurement of Economic Performance and Social Progress.

Tarrow, S. (2011). *Power in movement: Social movements and contentious politics* (3rd ed.). Cambridge University Press.

Westley, F., Zimmerman, B., & Patton, M. Q. (2011). *Getting to maybe: How the world is changed*. Vintage Canada.

Young, I. M. (1990). *Justice and the politics of difference*. Princeton University Press.

Chapter 4
Digital Mechanisms

With digital mechanisms we refer to the specific ways in which digital technologies create value within purpose ecosystems. Digital mechanisms are instruments that amplify, extend, and strengthen the collective action of actors who already share values and a vision of transformative change. When purpose is genuine and the enabling conditions of the territory are adequate, digital technologies can make a purpose ecosystem significantly more effective. When purpose is absent or weak, the most sophisticated digital infrastructure produces nothing more than a well-connected network without direction.

This chapter examines how digital mechanisms work in purpose ecosystems, what functions they serve, how they manifest differently depending on whether the ecosystem has emerged organically or has been initiated by policy, and what the fundamental relationship is between digital capability and Transformative Purpose.

The chapter proceeds in three sections. Section 4.1 develops a functional understanding of what digital mechanisms do in purpose ecosystems, organized around coordination, transparency, and learning. It asks what problems digital tools solve for purpose ecosystems and how they solve them. Section 4.2 examines how the relationship between purpose ecosystems and digital mechanisms differs fundamentally depending on whether the ecosystem has emerged organically from a territory or has been initiated by institutional intervention. The two pathways are not merely different in degree. They reflect different underlying logics that shape what digital mechanisms are, how they arise, and what role they play in the ecosystem's life. Section 4.3 addresses the deepest and most important question about digital mechanisms in purpose ecosystems, that is, the relationship between digital capability and Transformative Purpose. It argues that digital mechanisms can amplify purpose but cannot substitute for it, and that the greatest risk in the digital transformation of purpose ecosystems is not technological failure but purpose displacement, the gradual substitution of digital performance for genuine transformative commitment.

M. Nicotra and M. Romano, *Purpose Ecosystems*, Contributions to Management Science, https://doi.org/10.1007/978-3-032-38382-2_4

4.1 Digitalization for Coordination, Transparency, and Learning

Purpose ecosystems face fundamental challenges that digital technologies are particularly well-suited to address. Digital technologies support coordination, bringing together diverse actors with complementary capabilities, aligning their activities around shared objectives, and managing the complexity of multi-stakeholder collaboration across organizational, sectoral, and geographic boundaries. Furthermore, digital technologies defend transparency, making visible the social and environmental consequences of ecosystem activities, enabling forms of accountability that were previously impossible, and demonstrating to communities, funders, and policymakers what the ecosystem is achieving. Finally, digital technologies support learning. They continuously improve ecosystem strategies in response to evidence about what is working and what is not, adapting to changing conditions, and building the adaptive capacity that enables long-term transformative impact. Each of these challenges has existed as long as multi-actor collaboration has existed. What digital technologies change is the scale, speed, and precision with which they can be addressed. The digitalization of purpose ecosystem activities requires attention to the absorptive capacity of ecosystem actors and to the micro-foundational processes through which digital tools are adopted and used (Romano & Nicotra, 2020; Scuotto et al., 2021; Scuotto et al., 2022).

Coordination is the most fundamental of these challenges. Bringing together universities, social enterprises, public institutions, civic organizations, community groups, and private firms around a shared transformative mission means solving coordination problems of extraordinary complexity. These actors speak different institutional languages, operate on different time horizons, and have different understandings of what success looks like. They are geographically dispersed, organizationally heterogeneous, and often have limited prior history of collaboration. Face-to-face meetings, formal agreements, hierarchical oversight are not sufficient in relation to the scale and complexity of purpose ecosystems. They are too slow, too costly. Digital coordination mechanisms address this challenge by reducing the transaction costs of multi-actor collaboration. Platforms connect actors with complementary capabilities and resources and allow geographically dispersed teams to coordinate in real time. Platforms make also relevant information available to all ecosystem actors, reducing the information asymmetries (Nicotra et al., 2024; Romano & Nicotra, 2020). Collaborative knowledge management systems capture and share the learning generated through ecosystem activities, reducing the duplication and knowledge loss that plague fragmented multi-actor systems.

The coordination function is most powerful when it enables forms of collaboration that would otherwise be simply impossible, not merely difficult. Industrial symbiosis platforms that match the waste streams of industrial producers with the resource needs of other producers are a good example. The coordination problem they solve (matching diverse waste producers and resource users across geographic and sectoral boundaries in real time) is simply not solvable through conventional coordination mechanisms. Labor market platforms that connect marginalized workers with employers in sectors where they are underrepresented solve coordination problems that institutional intermediaries alone cannot address at scale. Community resource-sharing platforms that enable households to pool assets and reduce consumption create forms of collaborative consumption that were not organizationally feasible before digital coordination infrastructure made them possible.

Digital technologies reduce the transaction costs of market exchange and collaborative coordination (Coase, 1937; Williamson, 1985), enabling organizational forms and new collaborative arrangements. Parker et al. (2016) explain how coordination platforms generate network value. Actually, they increase their value as more actors participate. Besides, Stiglitz (2002) and Rochet and Tirole (2003) argue that digital technologies reduce the information asymmetrics that prevent effective coordination among actors with different knowledge, capabilities, and interests. Moreover, the organizational theory of collaborative advantage (Huxham & Vangen, 2005) clarifies why multi-actor collaboration generates value beyond what individual actors can achieve alone. There are some conditions that enable value to be realized rather than dissipated through coordination.

Transparency is the second core function, and in some respects the most distinctive. Purpose ecosystems claim to generate social and environmental transformation. That claim must be substantiated, not merely asserted. Digital monitoring and reporting systems create the informational infrastructure needed for substantiated accountability. They make visible the environmental and social consequences of ecosystem activities, enable the detection of gaps between stated commitments and actual behaviors, and provide the evidence base that funders, policymakers, and communities need to assess whether the ecosystem is genuinely fulfilling its Transformative Purpose.

Environmental monitoring systems offer clear examples of what transparency can be reached in practice. Satellite platforms can detect deforestation at a scale and speed that ground-based monitoring cannot match. In this way, actors can accountable for commitments that were once nearly impossible to enforce. IoT-based emissions systems track the environmental performance of ecosystem actors in real time. This gives all participants a clear picture of the collective environmental footprint and flags where action is needed. Supply chain traceability tools, built on blockchain or other distributed ledger technologies, create permanent, tamper-proof records of how products and materials move from origin to end use. This allows consumers, investors, and regulators to verify claims about environmental and social performance that previously could not be checked.

Social transparency mechanisms are equally important for purpose ecosystems committed to genuine social transformation. Service accessibility dashboards that track in real time who is and who is not reaching ecosystem services make equity gaps visible before they become entrenched. Participatory monitoring platforms that enable community members to report their experiences of ecosystem activities provide a bottom-up accountability mechanism that complements top-down performance reporting. Algorithms that make it possible to see when AI-driven decisions produce unfair outcomes across different groups. Open impact reporting platforms that publish ecosystem performance data in accessible, comparable formats create public accountability, that is, more robust than any internal monitoring system.

The transparency function of digital mechanisms is not merely instrumental. It is constitutive of the ecosystem's identity and legitimacy. Purpose ecosystems claim a distinctive form of accountability that goes beyond financial reporting or regulatory compliance. It is the accountability to the communities whose transformation they aspire to generate, to the values that motivate their activities, and to the long-term systemic changes that justify their existence. Digital transparency mechanisms are the primary means through which this distinctive accountability can be made real rather than rhetorical. An ecosystem that claims Transformative Purpose but cannot demonstrate, through credible monitoring and reporting, that its activities are generating the changes it claims to generate, is an ecosystem whose claims cannot be distinguished from purpose-washing.

Learning is the third core function, and the one that most directly connects digital capability to the adaptive capacity that long-term transformative impact requires. Purpose ecosystems operate in complex, rapidly changing environments. Social needs evolve, environmental conditions shift, policy frameworks change, new actors enter the ecosystem, and established actors exit or change their priorities. An ecosystem that cannot learn from experience, adapt its strategies in response to evidence, and revise its understanding of what its Transformative Purpose requires in changing circumstances will gradually lose its relevance and its effectiveness, regardless of how well it was initially designed. Digital platforms that support knowledge-sharing across ecosystem boundaries have been shown to be particularly important in enabling the open innovation dynamics on which ecosystem learning depends (Scuotto et al., 2019a; Scuotto et al., 2019b).

Digital learning mechanisms create the informational infrastructure for adaptive management, that is, the continuous cycle of monitoring, analysis, reflection, and adaptation through which complex systems maintain their effectiveness over time (Pahl-Wostl, 2009; Williams & Brown, 2018). Data analytics platforms that synthesize monitoring data from multiple sources enable ecosystem actors to detect patterns, trends, and anomalies that would be invisible in any single data stream. Besides, machine learning systems process large volumes of data that can be behavioral, operational, and environmental. In this way they can identify the mechanisms through which ecosystem activities generate outcomes, enabling more targeted and effective intervention strategies. In addition, rapid prototyping and experimentation platforms allow ecosystems to test innovations on a small scale before full implementation; in this way, they develop adaptive decision-making and so limit the costs of failure.

The learning function is particularly valuable for purpose ecosystems because the knowledge needed for effective social and environmental transformation is not static. Actually, such knowledge is co-produced through the ecosystem's activities, embedded in the experiences of the members, and revised as new evidence accumulates. Digital knowledge management systems capture and share such distributed and experiential knowledge in a way that ecosystem can learn collectively rather than individually. Cumulative knowledge assets are created, benefiting all actors and persisting through the turnover of individual participants. Collaborative platforms generate insights that no single actor could produce. They enable diverse ecosystem actors to interpret data together, drawing on their different perspectives and institutional logics.

Coordination, transparency, and learning functions are synthesized in Table 4.1. However, they are not independent functions. They are deeply interconnected and mutually reinforcing. Coordination platforms generate data about ecosystem activities that feed the transparency function. Transparency systems produce performance evidence that feeds the learning function. Learning generates adaptations in strategy and activity that require new coordination. And improved coordination produces more effective activities that generate better outcomes, closing the virtuous cycle. Digital mechanisms are most valuable in purpose ecosystems not when each function is optimized in isolation, but when the three are integrated into coherent systems that enable the entire cycle of purpose collective action to operate more effectively.

These functions come with risks, and they deserve honest acknowledgment. The coordination function can generate platform dependency. Ecosystems that rely heavily on specific digital platforms become vulnerable to changes in ownership, pricing, or governance. The transparency function can generate perverse incentives. When accountability relies mainly on digital metrics, ecosystem actors may focus on what is easy to measure, losing sight of the real and deeper transformative impact. The learning function can generate algorithmic bias. Machine learning systems trained on historical data will tend to replicate the patterns embedded in that data, including the patterns of exclusion and inequality that purpose ecosystems exist to challenge. None of these risks undermines the fundamental value of digital mechanisms for purpose ecosystem functioning. They are challenges to be managed, not reasons to avoid digitalization.

Table 4.1 Core functions of digital mechanisms in purpose ecosystems

Function	Core challenge addressed	Key digital tools	Primary risk
Coordination	Transaction costs of multi-actor collaboration across organizational and geographic boundaries	Matching platforms, shared project environments, open data platforms, collaborative knowledge systems	Platform dependency and lock-in
Transparency	Making social and environmental consequences of ecosystem activities visible and accountable	Environmental monitoring systems, supply chain traceability, service accessibility dashboards, open impact reporting	Indicator fixation and metric gaming
Learning	Adaptive management and continuous improvement in complex, changing environments	Data analytics, machine learning, rapid prototyping platforms, collaborative sense-making tools	Algorithmic bias and data quality failures

Source Authors' own elaboration

4.2 Two Pathways to Digital Integration

Digital mechanisms differ profoundly depending on whether the ecosystem has emerged organically from the interaction of actors who share values and purpose, or has been initiated and supported by institutional intervention. This is not merely a difference in the maturity or sophistication of digital infrastructure. It is a difference in the fundamental logic through which digital mechanisms relate to the ecosystem's purpose and collective action.

In organically emergent ecosystems, digital mechanisms are not prerequisites for ecosystem formation. They are instruments that arise in response to specific needs that actors experience as the ecosystem develops. The sequence matters. Purpose comes first, when actors who share values and a vision of change begin to collaborate, informally and relationally, before any digital infrastructure exists. As the collaboration deepens and scales, specific coordination problems emerge that actors recognize as limiting their effectiveness. Digital tools are then adopted, adapted, or created to address those specific problems. A community of social entrepreneurs working on circular economy transitions might begin with informal exchanges, then develop a shared database of local material flows, then build a matching platform as the number of actors and transactions grows beyond what informal coordination can manage. A network of civic organizations working on social transformation in a territory might begin with WhatsApp groups and shared documents, then develop more sophisticated systems as the network grows. The tools follow the needs.

This emergent pathway has several important characteristics. Digital tools in emergent ecosystems tend to be lightweight, modular, and closely tailored to the specific needs of the actors who use them. They are adopted because actors find them useful and they evolve incrementally as the ecosystem's needs change, rather than being designed comprehensively in advance (Baldwin & Woodard, 2009; Tiwana et al., 2010). And they are governed informally, through the same relational norms and shared values that govern the rest of the ecosystem's activities. Effective digital mechanisms in emergent ecosystems grow with the ecosystem. They start simple and then become more sophisticated with the development of needs and capabilities (Nambisan et al., 2019; Vial, 2019).

Even technically sophisticated tools, if imposed from outside, will not be embraced if people see them as ineffective for their specific needs (Davis, 1989; Rogers, 2003). The same applies to tools that assume a level of organizational maturity or digital competence that actors do not have yet. Rather than opening participation, they create new barriers. Tools that require comprehensive data sharing before actors have developed the relational trust needed to share sensitive information will generate resistance rather than transparency.

The open-source and commons-based peer production literature (Benkler, 2006; Bollier, 2014; Ostrom, 1990) provides important insights into how digital mechanisms can be governed in ways that are compatible with the distributed agency and shared values of emergent purpose ecosystems. Open-source software platforms, shared digital commons, and cooperatively governed data systems are digital infrastructure forms that align with the non-hierarchical, purpose-driven character of emergent ecosystems. They distribute control among users rather than concentrating it in a single operator, they enable community participation in the governance of shared digital resources, and they create the transparency and accountability that purpose ecosystems require without introducing the dependency risks that proprietary platforms generate. For emergent purpose ecosystems, these commons-based digital forms are not merely technically superior alternatives to proprietary platforms. They are expressions in digital form of the same values that define the ecosystem's identity.

The concept of digital social innovation, developed by Nesta (2012) and Mulgan (2019), describes a form of digital mechanism development that is particularly characteristic of emergent purpose ecosystems. It refers to the co-creation of digital tools with and by the communities they are intended to serve, drawing on local knowledge, lived experience, and community relationships that external technology developers lack. Digital social innovation treats communities not as users of pre-designed digital systems but as co-designers and co-owners of the digital infrastructure that shapes their collaborative life. This co-design approach generates tools that are more relevant, more trusted, and more likely to be genuinely adopted than externally designed systems. Moreover, it simultaneously builds the digital capabilities and confidence of community members who participate in the design process. For emergent purpose ecosystems, digital social innovation is not merely a methodology for tool development. It is a manifestation of the co-creation logic that is constitutive of purpose ecosystem identity.

In policy-initiated ecosystems, the relationship between digital mechanisms and ecosystem formation is fundamentally different. When institutional actors create purpose ecosystems through deliberate policy intervention, they typically also design and install the digital infrastructure intended to support the ecosystem's coordination, transparency, and learning functions. Digital mechanisms in policy-initiated ecosystems are planned and implemented as part of the ecosystem's formal structure, often before the collaborative relationships, shared values, and purpose engagement of ecosystem actors have fully developed. This reversal of the sequence creates opportunities and risks that are distinctive to the policy-initiated pathway.

The opportunities are significant. Policy-initiated ecosystems can deploy digital infrastructure at a scale and sophistication that emergent ecosystems, building incrementally from modest beginnings, may take years to achieve. Shared monitoring platforms, interoperable data systems, and collaborative governance tools can be designed comprehensively from the outset, creating from the beginning the informational infrastructure that enables ecosystem-wide transparency and collective learning. Formal data governance frameworks, developed as part of the ecosystem's institutional design, can address the privacy, sovereignty, and equity concerns that arise in shared data systems before they become sources of conflict or exclusion. And the legitimacy of institutional backing can accelerate the adoption of digital tools by actors who might otherwise be reluctant to share data or adopt new systems.

The risks are equally significant, and in some respects more fundamental. Digital infrastructure designed in advance of genuine collaborative relationships may not fit the actual coordination needs that actors experience once the ecosystem begins to operate (Dattée et al., 2018). Sophisticated transparency systems may generate performance reporting requirements that consume the time and energy of actors who should be focused on purpose activities rather than compliance (Power, 1999). Learning systems that presuppose organizational capabilities for data analysis and interpretation that many ecosystem actors do not possess may generate data without generating learning. And most importantly, the presence of comprehensive digital infrastructure from the outset may create the illusion of a functioning purpose ecosystem before the authentic shared purpose, relational trust, and genuine collaborative commitment that distinguish a real purpose ecosystem from a formally structured network have developed (Emerson & Nabatchi, 2015).

The most consequential risk is what we might call digital purpose substitution, the replacement of genuine transformative commitment with digital performance. When ecosystem actors are evaluated primarily through digital indicators, when accountability is demonstrated primarily through data dashboards, and when the ecosystem's legitimacy is constructed primarily through digital reporting rather than through demonstrated community impact, the digital infrastructure begins to serve its own perpetuation rather than the Transformative Purpose it was designed to support. Actors optimize for indicators rather than for impact. Reporting systems generate data without generating learning. Transparency mechanisms can create the appearance of accountability without the substance. This can be seen as purpose-washing where digital sophistication becomes a mask for the absence of genuine engagement with purpose.

To prevent purpose-washing in policy-initiated ecosystems, it is needed that digital tools are installed gradually, in response to coordination needs that emerge over time. In addition, governance frameworks should be designed with genuine community involvement, not just with technical experts and institutional stakeholders. In this perspective, digital mechanisms should be evaluated in the way they genuinely serve the ecosystem's Transformative Purpose.

The comparison between the two pathways reveals a fundamental asymmetry. In emergent ecosystems, purpose comes first and digital mechanisms develop in its service. The ecosystem's digital infrastructure is shaped by and accountable to the purposes that motivate its actors. In policy-initiated ecosystems, digital infrastructure often comes first and purpose must develop within the constraints and opportunities that the infrastructure creates. The ecosystem's purposes are partly shaped by the digital systems designed to measure and report them (Mazzucato, 2018; Schot & Steinmueller, 2018). This asymmetry does not make policy-initiated ecosystems inherently less purpose-driven than emergent ones. But it does create a distinctive governance challenge that is ensuring that digital infrastructure serves rather than shapes the ecosystem's evolving purpose, and that the metrics through which digital systems evaluate ecosystem performance reflect the genuine values of the community rather than the convenience of the institutions that designed the measurement systems (Winner, 1980; Zuboff, 2019).

Box 4.1 Two Pathways to Digital Mechanisms

In emergent ecosystems, purpose comes first and digital tools follow. Tools arise in response to specific needs. Development is incremental and organic. Governance is informal and relational. Risk: digital tools may not develop quickly enough to support scaling. Opportunity: tools are closely aligned with genuine actor needs and values.

In policy-initiated ecosystems, digital infrastructure is planned and installed as part of formal ecosystem design. Tools are comprehensive from the outset. Governance is formal and contractual. Risk is related to the fact that digital infrastructure may substitute for genuine purpose engagement. Opportunity derives from shared infrastructure at scale from the beginning.

In both cases, the fundamental principle is that digital mechanisms serve purpose. They do not generate it, and they cannot substitute for it. The quality of the ecosystem's transformative commitment is not visible in its digital infrastructure, however sophisticated. It is visible in the quality of its collaborative relationships, the genuineness of its community engagement, and the evidence of systemic change in its territory.

Spotlight 4.1 Decidim and the Politics of Open Digital Infrastructure

In 2016, the city of Barcelona needed a digital tool capable of enabling tens of thousands of citizens to genuinely participate in the development of a municipal action plan, contributing proposals, commenting on ideas, voting on priorities, and tracking the implementation of commitments. The city government, then led by Barcelona en Comú (a citizens' political movement that had itself emerged from the participatory culture of the 15-M social movement) commissioned the development of a new platform. What emerged was the most important examples available of how digital mechanisms can serve a Transformative Purpose without substituting for it.

Decidim, which means "we decide" in Catalan, was developed as free, open-source software from the beginning. This was not a secondary technical decision. It was a constitutive expression of the platform's purpose. A participatory democracy tool that was proprietary or controlled by a private vendor would embody a contradiction between its form and its function. By making the code public, the development community established that the infrastructure for democratic participation should be a common, owned by no one, and available to everyone. The Decidim Association, founded in 2019 to govern the platform's development, is a multi-stakeholder organization that includes developers, civil society organizations, researchers, and public institutions from across the world.

What makes Decidim an exemplary case for the argument of this chapter is the precision with which it illustrates the amplification principle. Digital mechanisms amplify what already exists. In Barcelona, the platform amplified a participatory culture that was already present in the city's social movements, neighborhood associations, and civic organizations. The 70,000 Barcelona residents who participated in the first municipal action plan process through Decidim did not create their desire for democratic participation because the platform made it easy. They had been wanting genuine participation for years; Decidim gave them a credible and effective instrument through which to exercise it.

By 2024, Decidim was being used in over 400 organizations across more than 40 countries, including cities from Helsinki to New York to Mexico City, national governments, universities, and civil society organizations. Each deployment has adapted the platform's capabilities to its specific context. It also maintains the core commitment to open, transparent, and participatory governance. The critical lesson Decidim offers is about the relationship between technical design and political values. Every design choice in a digital platform embeds assumptions about who has power, whose voice counts, and what transparency means. Decidim's developers made these assumptions explicit and contestable, building the platform's governance principles into its technical architecture in ways that can be audited, challenged, and revised by the communities that use it. This is the operationalization of the amplification principle: building digital infrastructure that serves and extends the democratic values of the communities it is designed to empower.

4.3 Amplification Without Substitution

Digital mechanisms can amplify Transformative Purpose, but they cannot generate it or substitute for it. This is not a technological limitation that might be overcome by more sophisticated systems. It is a logical necessity that follows from what Transformative Purpose is. Purpose emerges from the interaction of actors who genuinely share values, concerns, and a vision of the change they want to create. It is co-constructed through dialogue, joint action, and shared learning. It is sustained by the relational trust, mutual commitment, and collective identity that develop among actors through their collaborative experience. None of these things, not shared values, not genuine co-construction, not relational trust, not collective identity can be generated by a digital system, however sophisticated. They are human and relational phenomena that digital tools can support but cannot create.

What digital mechanisms can do is make a purpose ecosystem that already has genuine Transformative Purpose significantly more effective. Coordination tools that connect actors more efficiently allow a genuine collaborative commitment to translate into more comprehensive and impactful collective action. Transparency tools that make the social and environmental consequences of ecosystem activities visible allow a genuine accountability commitment to be demonstrated rather than merely asserted. Learning tools that process evidence from ecosystem activities enable actors who are genuinely committed to improvement to learn more rapidly and adapt more effectively. In each case, the digital mechanism amplifies a commitment that already exists. It extends its reach, increases its precision, and accelerates its effects. But it cannot supply the commitment in the first place.

An analogy from organizational theory helps to clarify this. Management information systems amplify the strategic capabilities of organizations with clear strategies and capable leadership, enabling them to execute their strategies more precisely and learn more rapidly. But they cannot supply the strategic clarity or leadership capability that organizations lack. Organizations without clear strategies and capable leadership that invest heavily in management information systems typically become more efficiently confused, not more strategically effective. The same logic applies to digital mechanisms in purpose ecosystems. They amplify the capabilities of ecosystems with genuine Transformative Purpose, but they cannot supply the purpose that ecosystems lack.

This principle has important implications for how digital mechanisms should be evaluated in purpose ecosystems. The important question is whether the digital infrastructure is genuinely serving the ecosystem's Transformative Purpose. Are coordination tools enabling more inclusive and effective collaborative action? Are transparency tools generating accountability that changes behavior in purpose directions? Are learning tools producing adaptations that bring ecosystem activities into closer alignment with their transformative objectives? Digital mechanisms that score well on the first question but poorly on the second are digital infrastructure in search of a purpose, which is precisely the condition that makes purpose-washing possible.

Purposive displacement is the technical name for this drift, and it deserves sustained analytical attention. It occurs when the means of ecosystem activity, and so digital infrastructure, reporting systems, performance metrics, gradually become ends in themselves, displacing the Transformative Purposes they were designed to serve (Merton, 1957; Selznick, 1957). This is not a rare or pathological phenomenon. It is a predictable consequence of the institutional pressures that all organizations and ecosystems face.

On one hand, funders demand performance data which pushes ecosystems to optimize for whatever gets measured (Espeland & Sauder, 2007; Power, 1999). On the other hand, regulators require compliance documentation, which drives investment into reporting systems. Moreover, evaluators treat digital technologies as a signal of ecosystem quality, and this creates pressure to deploy digital tools regardless of whether they serve any genuine transformative need. Over time, these pressures can produce ecosystems that look digitally sophisticated, formally accountable, well-documented but that have quietly lost the authentic purpose engagement and community rootedness. This is coherent with theories on institutional isomorphism (DiMaggio & Powell, 1983; Suchman, 1995; Scott, 1995). It describes a familiar tendency of organizations that end up adopting the structures, practices, and symbols that their institutional environment rewards, regardless of whether any of it works.

For purpose ecosystems, the relevant institutional environment increasingly includes funders, evaluators, and policymakers who value digital sophistication as a marker of ecosystem quality and seriousness. The isomorphic pressure to adopt and display digital capabilities can drive purpose ecosystems toward digital investments that serve institutional legitimacy rather than Transformative Purpose, gradually displacing the latter with the former.

Resisting this displacement requires ecosystems to maintain a clear and explicit distinction between digital mechanisms as instruments and Transformative Purpose as the ultimate criterion of ecosystem value. This distinction must be embedded in the ecosystem's evaluation and accountability systems. Performance frameworks that privilege quantitative digital indicators over qualitative and participatory evidence of transformative impact will systematically accelerate purposive displacement. It must be embedded in the ecosystem's governance processes and in the ecosystem's culture. Decision-making bodies that rely primarily on digital dashboards for their understanding of ecosystem performance will gradually lose touch with the lived experience of the communities whose transformation they are responsible for. At the same time, communities of practice that celebrate digital sophistication as a primary marker of ecosystem quality will create incentives for actors to invest in digital performance rather than in purpose engagement.

The relationship between digital mechanisms' social transformation and environmental sustainability deserves specific attention. Digital mechanisms can amplify both dimensions. However, the amplification dynamics are different, and the risks of substitution are distributed unevenly.

For environmental sustainability, digital amplification is particularly powerful and the risks of substitution are relatively manageable. Environmental monitoring and reporting systems generate credible, quantifiable evidence. Environmental performance can be verified through specific digital systems like satellite imagery, emissions measurements, material flow analysis, and biodiversity monitoring. Such systems create robust and reliable accountability, that is, structurally resistant to greenwashing. The main risk in the environmental domain is that digital monitoring systems create incentives to focus on indicators easily measured, such as energy efficiency or carbon emissions, instead of more complex and ecologically significant dimensions of sustainability, such as biodiversity, soil health, and water quality, harder to monitor digitally.

For social transformation, the risks are more difficult to manage. Social transformation is a relational and cultural phenomenon. It involves changes in the quality of human relationships, the distribution of power, the cultural norms of communities, and the institutional practices. These dimensions are not easily captured by digital indicators, and the attempt to do so systematically biases accountability systems toward the more measurable and less important dimensions of social change. Digital transparency mechanisms that report on service delivery metrics, participation counts, and income changes capture important but partial dimensions of social transformation, systematically undervaluing the changes in community agency, cultural recognition, and institutional accountability that represent the deepest and most durable contributions that purpose ecosystems can make.

This analysis points toward a principle of proportionate digitalization. Digital mechanisms should be deployed in proportion to their capacity to genuinely serve the ecosystem's Transformative Purpose, and the dimensions of purpose that are least amenable to digital measurement should receive the most deliberate protection from the accountability pressures that digital systems generate. Environmental sustainability monitoring can and should be extensively digitalized, because digital monitoring genuinely strengthens environmental accountability without significant risk of displacing the values it is meant to serve. Social transformation accountability should be partially digitalized, but the digital component should always be complemented by participatory, qualitative, and community-centered forms of evidence that capture the dimensions of change that digital systems cannot measure. And purpose itself (the shared commitment that motivates ecosystem actors and gives meaning to their activities) should never be made subject to digital performance management. It is the dimension of ecosystem life that most fundamentally requires human judgment, relational sensitivity, and collective deliberation.

The relationship between digital mechanisms and Transformative Purpose is not static. It evolves as the ecosystem develops, as digital technologies change, and as the community's understanding of its Transformative Purpose deepens. In the early stages of a purpose ecosystem's development, digital mechanisms play a limited role. The ecosystem's energy is invested in building the relational trust, developing the shared language, and discovering the common values that constitute authentic purpose. As the ecosystem matures and its collaborative activities scale, digital mechanisms become more important as coordination and transparency instruments. And as the ecosystem engages with policy systems, funders, and external evaluators, digital reporting and accountability systems become more salient. Throughout this evolution, the fundamental principle remains constant, that is, digital mechanisms serve purpose. They extend it, amplify it, and make it visible. But they do not create it, and they cannot save an ecosystem that has lost it.

Box 4.2 The Amplification Principle for Digital Mechanisms

Digital mechanisms amplify what already exists. In ecosystems with genuine Transformative Purpose, they make that purpose more effective. In ecosystems without genuine Transformative Purpose, they make the absence of purpose more efficiently invisible. The quality of digital mechanisms is therefore a necessary but not sufficient indicator of ecosystem quality. The ultimate test is always the quality of the ecosystem's transformative commitment and the evidence of systemic change in its territory.

Therefore, digital mechanisms can make coordination among diverse actors faster, cheaper, and more comprehensive. They create transparency about the social and environmental consequences of ecosystem activities. Moreover, they enable learning from evidence and continuous improvement of ecosystem strategies. Finally, they extend the reach of ecosystem activities to actors and communities that would otherwise be difficult to engage, and demonstrate accountability to funders, policymakers, and communities.

However, digital mechanisms cannot generate the shared values that motivate genuine purpose ecosystem actors. Actually, they are not able to create the relational trust that sustains authentic multi-actor collaboration or to define or sustain the Transformative Purpose that gives the ecosystem its identity and direction. Therefore, they cannot substitute for the community co-creation and participatory governance that give purpose ecosystems their legitimacy or guarantee that reported performance reflects genuine transformative impact.

Spotlight 4.2 Missions, Data, and the Digital Backbone of Impact Innovation in Sweden

In 2023, Sweden launched Impact Innovation, a joint initiative of three national agencies (Vinnova, the Swedish Energy Agency, and Formas) designed to become Sweden's primary innovation investment for accelerating the transition to a sustainable society. Organized around explicit societal missions (a fossil-free and circular industry, sustainable water management, a reformed public sector, and smart and sustainable built environments) Impact Innovation had activated over 400 projects across the country by early 2026, engaging actors from industry, academia, the public sector, and civil society, and investing over 1.6 billion Swedish kronor in the transformation of Sweden.

What makes Impact Innovation instructive for this chapter is not primarily its scale but its approach to digital mechanisms. The program was designed from the outset around the principle that digital infrastructure should serve the missions' Transformative Purposes, not substitute for them. The three funding agencies brought to Impact Innovation decades of experience with Strategic Innovation Programmes, and one of the most important lessons from that experience was that digital coordination tools developed in advance of genuine collaborative relationships tend to create compliance infrastructure rather than genuine learning. Impact Innovation's digital architecture was therefore designed to follow the development of actor relationships, not to precede them.

The transparency function of digital mechanisms is particularly significant given the OECD's explicit study of Impact Innovation's governance, published in 2024, as a model of "ecosystem-driven missions." The OECD found that Impact Innovation's open reporting practices (making program activities, results, and governance decisions accessible to all stakeholders) create accountability, that is, substantively different from the accountability of conventional government programs. Because the data is publicly available and the program offices are genuinely actor-driven rather than government-directed, civil society organizations and academic researchers can independently assess whether program activities are genuinely contributing to mission objectives, creating a form of distributed oversight that formal evaluation frameworks cannot replicate.

The learning function of Impact Innovation's digital infrastructure reflects that digital learning systems should generate adaptation, not merely documentation. When data indicates that an approach is not working, the expectation is not that the program will document the failure and continue as planned, but that it will genuinely adapt. The digital infrastructure makes this adaptation visible and accountable; program offices must show how their data has influenced their strategic choices, creating a feedback loop between digital measurement and purpose action that operationalizes the learning function at scale. Impact Innovation demonstrates that sophisticated digital mechanisms in the service of genuine missions can generate a form of accountability that neither conventional government reporting nor market discipline can produce.

References

Akerlof, G. A. (1970). The market for "lemons": Quality uncertainty and the market mechanism. *Quarterly Journal of Economics, 84*(3), 488–500.

Baldwin, C. Y., & Woodard, C. J. (2009). The architecture of platforms: A unified view. In A. Gawer (Ed.), *Platforms, markets and innovation* (pp. 19–44). Edward Elgar.

Benkler, Y. (2006). *The wealth of networks: How social production transforms markets and freedom.* Yale University Press.

Bollier, D. (2014). *Think like a commoner: A short introduction to the life of the commons.* New Society Publishers.

Buterin, V. (2014). *A next-generation smart contract and decentralized application platform.* Ethereum White Paper.

Catalini, C., & Gans, J. S. (2020). Some simple economics of the blockchain. *Communications of the ACM, 63*(7), 80–90.

Coase, R. H. (1937). The nature of the firm. *Economica, 4*(16), 386–405.

Dattée, B., Alexy, O., & Autio, E. (2018). Maneuvering in poor visibility: How firms play the ecosystem game when uncertainty is high. *Academy of Management Journal, 61*(2), 466–498.

Davis, F. D. (1989). Perceived usefulness, perceived ease of use, and user acceptance of information technology. *MIS Quarterly, 13*(3), 319–340.

DiMaggio, P. J., & Powell, W. W. (1983). The iron cage revisited: Institutional isomorphism and collective rationality in organizational fields. *American Sociological Review, 48*(2), 147–160.

Doshi-Velez, F., & Kim, B. (2017). Towards a rigorous science of interpretable machine learning. arXiv preprint arXiv:1702.08608.

Emerson, K., & Nabatchi, T. (2015). *Collaborative governance regimes.* Georgetown University Press.

Espeland, W. N., & Sauder, M. (2007). Rankings and reactivity: How public measures recreate social worlds. *American Journal of Sociology, 113*(1), 1–40.

Gawer, A., & Cusumano, M. A. (2014). Industry platforms and ecosystem innovation. *Journal of Product Innovation Management, 31*(3), 417–433.

Huxham, C., & Vangen, S. (2005). *Managing to collaborate: The theory and practice of collaborative advantage.* Routledge.

Jacobides, M. G., Cennamo, C., & Gawer, A. (2018). Towards a theory of ecosystems. *Strategic Management Journal, 39*(8), 2255–2276.

Mazzucato, M. (2018). *The value of everything: Making and taking in the global economy.* Allen Lane.

Merton, R. K. (1957). *Social theory and social structure.* Free Press.

Mulgan, G. (2019). *Social innovation: How societies find the power to change.* Policy Press.

Nambisan, S., Wright, M., & Feldman, M. (2019). The digital transformation of innovation and entrepreneurship. *Research Policy, 48*(8), 103773.

Nesta. (2012). *Digital social innovation: What it is and what we need to do next.* Nesta.

Nicotra, M., Munnia, A., Russo, F., & Zingale, A. (2024). Open data-driven ecosystems: An empirical validation. In J. A. Cunningham, M. Menter, C. O'Kane, & M. Romano (Eds.), *Research handbook on entrepreneurial ecosystems.* Edward Elgar.

Ostrom, E. (1990). *Governing the commons: The evolution of institutions for collective action.* Cambridge University Press.

Pahl-Wostl, C. (2009). A conceptual framework for analysing adaptive capacity and multi-level learning processes in resource governance regimes. *Global Environmental Change, 19*(3), 354–365.

Parker, G. G., Van Alstyne, M. W., & Choudary, S. P. (2016). *Platform revolution.* W. W. Norton.

Power, M. (1999). *The audit society: Rituals of verification.* Oxford University Press.

Rochet, J. C., & Tirole, J. (2003). Platform competition in two-sided markets. *Journal of the European Economic Association, 1*(4), 990–1029.

Rogers, E. M. (2003). *Diffusion of innovations* (5th ed.). Free Press.

Romano, M., & Nicotra, M. (2020). Technologies 4.0 towards digital transformation. In S. Bresciani, A. Ferraris, M. Romano, & G. Santoro (Eds.), *Digital transformation Management for Agile Organizations*. Emerald.

Schot, J., & Steinmueller, W. E. (2018). Three frames for innovation policy: R&D, systems of innovation and transformative change. *Research Policy, 47*(9), 1554–1567.

Scott, W. R. (1995). *Institutions and organizations*. SAGE.

Scuotto, V., Arrigo, E., Candelo, E., & Nicotra, M. (2019b). Ambidextrous innovation orientation effected by the digital transformation. *Business Process Management Journal, 26*(5), 1121–1140.

Scuotto, V., Magni, D., Paladino, R., & Nicotra, M. (2022). Triggering disruptive technology absorptive capacity by CIOs. *Technological Forecasting and Social Change, 174*, 121234.

Scuotto, V., Nicotra, M., Del Giudice, M., Krueger, N., & Gregori, G. L. (2021). A microfoundational perspective on SMEs' growth in the digital transformation era. *Journal of Business Research, 129*, 382–392.

Scuotto, V., Orlando, B., Cillo, V., Nicotra, M., & Di Gioia, L. (2019a). Uncovering the micro-foundations of knowledge sharing in open innovation partnerships. *Technological Forecasting and Social Change, 152*, 119906.

Stiglitz, J. E. (2002). Information and the change in the paradigm in economics. *American Economic Review, 92*(3), 460–501.

Suchman, M. C. (1995). Managing legitimacy: Strategic and institutional approaches. *Academy of Management Review, 20*(3), 571–610.

Tapscott, D., & Tapscott, A. (2016). *Blockchain revolution: How the technology behind bitcoin is changing money, business, and the world*. Portfolio/Penguin.

Tiwana, A. (2014). *Platform ecosystems: Aligning architecture, governance, and strategy*. Morgan Kaufmann.

Tiwana, A., Konsynski, B., & Bush, A. A. (2010). Research commentary: Platform evolution: Coevolution of platform architecture, governance, and environmental dynamics. *Information Systems Research, 21*(4), 675–687.

Vial, G. (2019). Understanding digital transformation: A review and a research agenda. *Journal of Strategic Information Systems, 28*(2), 118–144.

Williams, B. K., & Brown, E. D. (2018). Double-loop learning in adaptive management: The need, the challenge, and the opportunity. *Environmental Management, 62*(6), 995–1006.

Williamson, O. E. (1985). *The economic institutions of capitalism*. Free Press.

Winner, L. (1980). Do artifacts have politics? *Daedalus, 109*(1), 121–136.

Yoo, Y., Boland, R. J., Lyytinen, K., & Majchrzak, A. (2012). Organizing for innovation in the digitized world. *Organization Science, 23*(5), 1398–1408.

Zuboff, S. (2019). *The age of surveillance capitalism: The fight for a human future at the new frontier of power*. PublicAffairs.

Chapter 5
Governance Architecture

Governance Architecture consists of the rules, processes, accountability mechanisms, and institutional arrangements through which a purpose ecosystem makes collective decisions, maintains alignment with its shared purpose, and remains accountable to the communities it serves. Governance is what holds the other three components of the compass together. Without it, enabling conditions cannot be directed toward purpose ends, digital mechanisms cannot be held accountable to transformative values, and Transformative Purpose cannot be protected from the persistent pressures that erode it over time.

Yet, governance in purpose ecosystems is not what governance typically means in management scholarship. In conventional firms, governance refers to the structures through which shareholders exercise oversight over management. In public organizations, it refers to the structures through which elected officials exercise authority over public administrators. In purpose ecosystems, neither of these models applies. Purpose ecosystems are not hierarchically controlled. No single actor has the authority to direct the activities of all others. They are not governed by ownership rights or electoral mandates. They are collaborative systems whose coherence depends on the voluntary commitment of diverse actors to a shared purpose that none of them fully controls and all of them partially constitute.

The governance of ecosystems is extensively studied in the management literature. Entrepreneurial ecosystem scholars have documented how governance must simultaneously coordinate diverse actors, protect shared resources, and maintain the institutional legitimacy that attracts further participation (Autio & Thomas, 2019; Spigel, 2017; Stam, 2015). Innovation ecosystem research has shown how governance structures determine whether multi-actor alignment around a shared value proposition is sustained or progressively eroded by the diverging interests of participants (Adner, 2017; Jacobides et al., 2018). Platform ecosystem scholarship has demonstrated how governance rules shape who benefits from ecosystem membership and who is excluded (Gawer & Cusumano, 2014; Parker et al., 2016). What this

M. Nicotra and M. Romano, *Purpose Ecosystems*, Contributions to Management Science, https://doi.org/10.1007/978-3-032-38382-2_5

growing body of work shares is that governance of complex multi-actor systems is never merely technical. It is always also a problem of power, interest, and legitimacy.

The principal-agent problem in multi-actor ecosystems is particularly complex. In a dyadic relationship, a principal delegates authority to an agent and attempts to align the agent's interests through monitoring, incentives, and sanctions. In purpose ecosystems, this logic multiplies because every actor is simultaneously a principal in some relationships and an agent in others, and the ultimate principal, the community, is often the actor least represented in formal governance arrangements (Emerson et al., 2012; Provan & Kenis, 2008). Governing purpose ecosystems therefore requires not only mechanisms for aligning individual actor interests but structural arrangements capable of representing community voice, maintaining shared purpose, and redistributing power. This is what makes the governance challenge in purpose ecosystems more demanding than in other ecosystem types.

Governance in purpose ecosystems must coordinate without commanding, hold accountable without surveilling, maintain direction without prescribing, and protect purpose without ossifying it. How this challenge is met depends fundamentally on the nature of the ecosystem itself. Section 5.1 examines governance in organically emergent ecosystems, where governance arises from the bottom-up, is informal and relational, and grows with the ecosystem as its needs evolve. Section 5.2 examines governance in policy-initiated ecosystems, where formal structures are present from the outset and the central challenge is ensuring that those structures serve rather than supplant authentic purpose. Section 5.3 steps back from both cases to address the deeper theoretical questions on the nature of governance legitimacy without hierarchy, the role of power in systems that claim to share it equitably, and the fundamental paradox of structure in organizations whose most important asset, genuine Transformative Purpose, is precisely what formal structure most easily destroys.

5.1 Informal, Relational, and Adaptive Governance in Emergent Ecosystems

In organically emergent purpose ecosystems, governance does not precede the ecosystem; it emerges with it. As actors who share values and a vision of change begin to collaborate, they develop, through their collaborative experience, the informal norms, shared expectations, and relational accountability mechanisms that constitute governance in the absence of formal structures. Nobody designs this governance. It grows from the patterns of interaction, the commitments actors make to each other, and the reputational consequences of fulfilling or failing to fulfill those commitments. It is governance by practice rather than governance by structure.

The concept of practice-based governance, drawing on the organizational theory of practice (Bourdieu, 1977; Giddens, 1984; Schatzki et al., 2001) and the situated learning literature (Lave & Wenger, 1991; Wenger, 1998), provides an appropriate theoretical framework for understanding how governance emerges in organically developed purpose ecosystems. Social structures are not external constraints on action but are continuously reproduced and sometimes transformed through the very practices they constrain and enable. For purpose ecosystems, governance norms and expectations emerge from and are continuously reproduced through the repeated practices of collaboration, the meetings, joint projects, shared decision-making processes, and collective reflections through which ecosystem actors develop a shared understanding of how things should be done, what behaviors are acceptable, and what commitments can be relied upon.

The sociology of trust provides the most developed theoretical account of the relational foundations of governance in emergent purpose ecosystems. Luhmann (1979) distinguished between trust in persons (based on accumulated experience of specific individuals) and trust in systems (based on confidence in institutional arrangements that make trustworthy behavior possible independently of personal relationships). In the early stages of emergent purpose ecosystems, governance depends primarily on person-based trust. Actors cooperate because in prior interactions they have developed shared experiences, demonstrated commitments, and accumulated reputational knowledge. As ecosystems grow, they develop more system governance mechanisms. We refer to shared norms, procedural expectations, and accountability structures that enable actors to cooperate with individuals they do not personally know on the basis of confidence in the ecosystem's governance culture.

The concept of embeddedness adds a complementary dimension (Granovetter, 1985). The economic and organizational action is shaped by the social relationships in which actors are embedded. In emergent purpose ecosystems, governance is deeply embedded in the relational fabric of the ecosystem. Actors behave in accordance with shared commitments not primarily because they fear formal sanctions but because they are embedded in networks of relationships in which their reputation, trustworthiness, and purpose commitment are continuously observed and evaluated by others whose good opinion they value. This embeddedness makes relational governance robust and fragile. Robust means that it generates genuine commitment that formal mechanisms cannot replicate. Fragile means that it depends on the continued density and quality of the relational networks.

Governance in emergent purpose ecosystems is informal. The rules and norms that govern ecosystem are not codified in formal procedures. They exist in the shared understanding of actors who have developed, through collaborative experience, a sense of what the ecosystem stands for, how decisions are made, what behaviors are acceptable, and what commitments can be relied upon. This informality is not a deficiency. It is an asset that enables governance to be flexible, responsive, and aligned with the genuine values of actors involved rather than with the procedural requirements of formal accountability systems.

Furthermore, governance in emergent ecosystems is relational. The most important governance mechanisms are not formal rules or accountability procedures but the practices of mutual respect, honest communication, and equitable benefit-sharing that build and maintain the trust relationships on which collective action depends. When governance in an emergent ecosystem fails, the failure is typically relational. This means that trust has been damaged by perceived betrayals of commitment, power differentials have undermined the quality of collaborative relationships, or the departure of key relationship holders has left gaps in the relational fabric that formal governance structures cannot fill. Restoring governance capacity in these situations requires restoring the relational foundations of collective action, which takes time and requires honest acknowledgment of what went wrong.

Lastly, governance in emergent ecosystems is adaptive because it evolves continuously as the ecosystem learns from its experience, encounters new challenges, and incorporates new actors. The governance norms and practices of an emergent ecosystem are not fixed. They are continuously reproduced and sometimes revised through the very practices they govern. When actors find that existing norms are no longer producing the outcomes they aim for, they adjust their practices and gradually build new shared expectations. This capacity to evolve can be seen as the ecosystem's most valuable governance asset. It allows the ecosystem to be relevant and responsive as circumstances change, without formal revision processes typical of more institutional systems.

To understand how non-hierarchical governance systems can be accountable and sustainable, we can refer to the real-world evidence of Elinor Ostrom's research (1990, 2010). Her eight design principles, developed from systematic comparative analysis of thousands of commons governance systems across diverse institutional and cultural contexts, identify the structural conditions under which non-hierarchical governance succeeds. These are clearly defined boundaries, governance rules that match local conditions, collective choice arrangements that enable those most affected to participate in rule-making, monitoring systems that enable mutual accountability, graduated sanctioning mechanisms, conflict resolution mechanisms, recognition of rights to organize, and nested governance structures for larger-scale commons. These principles are directly applicable to emergent purpose ecosystem governance, with adaptations for the normative dimension that natural resource commons do not share. This means maintaining the alignment of collective activities with a shared Transformative Purpose that is continuously evolving and vulnerable to capture.

The community of practice concept (Lave & Wenger, 1991; Wenger, 1998) is relevant to understand how governance norms are maintained and transmitted in emergent purpose ecosystems. Communities of practice develop routines, ways of doing things, stories, symbols, and concepts that constitute the living culture of the community. For emergent purpose ecosystems, the governance norms and practices that constitute their informal governance architecture are a key element of this shared repertoire. They are transmitted to new members through participation, demonstrated through practice, and revised through the community's ongoing dialogue about what it means to be a genuinely purpose ecosystem actor.

The challenge of scaling governance in emergent ecosystems is one of the most important practical challenges that purpose ecosystem leaders face. As ecosystems grow and actors become numerous, geographic scope larger, and activity diversified, the relational governance mechanisms that work well in small communities become increasingly strained (Ahuja, 2000; Provan & Kenis, 2008). Face-to-face relationships cannot be maintained with all ecosystem actors when the ecosystem exceeds a certain scale. Reputational accountability mechanisms lose effectiveness when actors are embedded in dense local networks but have limited visibility into distant or unfamiliar parts of the ecosystem. Addressing the scale challenge without sacrificing the relational and adaptive character of emergent governance requires nested governance, that is, a multi-level architecture in which small, relationally governed sub-communities are embedded within larger governance structures that provide strategic coherence and territorial legitimacy, without requiring hierarchical control from any single center (Ansell & Gash, 2008; Ostrom, 2010).

Backbone organizations, identified by Kania and Kramer (2011) as crucial for supporting multi-actor collaborations, are particularly important in emergent purpose ecosystems during the vulnerable transitions between small relational governance and larger institutionalized governance. They can facilitate the deliberative processes through which governance norms are developed and revised, provide the administrative capacity needed for ecosystem-wide coordination, and maintain institutional memory when key actors exit (Emerson et al., 2012; Huxham & Vangen, 2005). But they also create governance risks; they can accumulate disproportionate influence over ecosystem direction, develop organizational interests that diverge from the ecosystem's shared purpose, and create governance dependencies that undermine the distributed agency constitutive of purpose ecosystem identity.

Conflict in emergent ecosystems is inevitable and, when managed well, is productive. All multi-actor collaborations generate disagreements about priorities, resource allocation, and the interpretation of shared commitments (Gray, 1989; Innes & Booher, 2010). In emergent ecosystems, neither hierarchical authority nor exit through market mechanisms is available for conflict resolution. Governance must rely on deliberation, mediation, and the re-invocation of shared purpose (Habermas, 1984; Susskind & Cruikshank, 1987). The most damaging conflicts are suppressed rather than resolved. When actors avoid difficult conversations for fear of straining relationships, tensions do not disappear. They simply build up quietly beneath the surface, until they eventually erupt into governance crises (Argyris & Schön, 1978). Building governance cultures that treat conflict as a normal and productive dimension of collaborative life is therefore an essential governance design challenge.

Table 5.1 synthesizes the main characteristics of governance in emergent purpose ecosystems.

Spotlight 5.1 Bologna's Collaborative Governance of Urban Spaces

In 2014, the City of Bologna introduced what has become one of the most studied and replicated governance innovations in contemporary urban administration, the Regulation for the Care and Regeneration of Urban Commons. The regulation created a formal

Table 5.1 Characteristics of governance in emergent purpose ecosystems

Governance feature	Character in emergent ecosystems	Primary strength	Primary vulnerability
Rules and norms	Informal, practice-based, culturally embedded	Flexible, genuinely aligned with actor values	Depends on key carriers of culture
Accountability	Relational, reputational, network-embedded	Robust where trust is high and networks dense	Weakens as ecosystem scales
Decision-making	Consensus-oriented, deliberative, distributed	Generates genuine commitment and legitimacy	Slow, vulnerable to capture by vocal actors
Conflict resolution	Dialogue-based, mediated, purpose-referenced	Preserves relationships while addressing disagreements	Requires willingness to confront tensions openly
Adaptation	Continuous, practice-driven, culturally embedded	Rapid, responsive to genuine ecosystem experience	Difficult to coordinate at scale

Source Authors' own elaboration

legal framework through which citizens and civic organizations could enter into "collaboration pacts" with the municipal government to care for, manage, and regenerate shared urban spaces and services: parks, abandoned buildings, public libraries, school gardens, and digital services. By 2020, Bologna had signed more than 500 collaboration pacts, involving thousands of citizens, hundreds of civic organizations, and virtually every neighborhood in the city.

Reading the Bologna experience through the governance framework of this chapter reveals the anatomy of emergent governance in a purpose ecosystem with unusual clarity. The Regulation does not create governance from above. It recognizes and legitimizes governance practices that were already developing from below. The collaboration pacts are not contracts in the conventional legal sense, specifying obligations and penalties. They are relationships, expressions of mutual commitment between civic actors who want to contribute to their community and a municipal government that recognizes the legitimacy of that contribution. The Labsus organization, which developed the conceptual framework for the regulation and has supported its diffusion to over 270 Italian municipalities since 2014, functions precisely as the backbone organization identified in this chapter, a support structure that enables governance without directing it.

The governance of the Bologna commons illustrates the relational, adaptive, and practice-based character of emergent governance in ways that formal governance theory often misses. The collaboration pacts evolve as the relationships between civic actors and the municipality develop. Initial pacts are typically modest and exploratory; longer-standing relationships generate more ambitious collaborative activities. Governance norms develop through practice. Civic actors learn, through their experience of managing shared spaces, what works and what does not, and these lessons diffuse informally through the networks involved in the commons governance system.

> The critical governance lesson that Bologna offers is about the relationship between formal institutional recognition and informal relational governance. The Regulation did not generate the civic energy that produced the collaboration pacts. That energy was already present in Bologna's rich civic culture, with its dense networks of associations, cooperatives, and neighborhood groups. What the Regulation did was create a legitimate institutional channel through which that energy could be directed toward the care of shared spaces, transforming informal civic activity into a recognized and supported form of governance participation. This is governance as recognition. Not defining purpose, but creating the conditions within which genuine civic purpose can express itself effectively and sustainably.

5.2 Formal Structures, Accountability, and Purpose Integrity in Policy-Initiated Ecosystems

Policy-initiated purpose ecosystems, and so those created and supported by formal policy instruments including national innovation programs, European structural funds, and territorial development strategies, present a fundamentally different governance challenge from organically emergent ecosystems. They have formal governance structures from the outset, that is, governance documents, designated leadership bodies, contractual accountability relationships, reporting obligations, and institutional oversight mechanisms. They operate within the institutional constraints of the policy environments that created them. And they face external funders and policy principals who require demonstrated performance against predetermined objectives.

These formal governance structures create advantages and risks. The advantages are significant, such as clarity about decision-making authority and accountability relationships, institutional legitimacy that enables access to public resources and credible commitments to partners and communities, organizational infrastructure that sustains ecosystem activities beyond what informal relational governance can support, and accountability mechanisms that protect against misuse of public resources (Provan & Kenis, 2008; Skelcher & Smith, 2015). The risks are equally significant. The most important is governance substitution, the replacement of genuine purpose engagement with formal compliance performance. When governance is assessed primarily through compliance, when accountability is demonstrated primarily through reporting to funders rather than to communities, and when the ecosystem's legitimacy is constructed primarily through formal governance credentials rather than through genuine community rootedness, formal governance becomes a performance rather than a practice (Meyer & Rowan, 1977; Suchman, 1995).

The concept of organizational decoupling, developed by Meyer and Rowan (1977) and extensively applied in institutional theory, provides the theoretical framework for understanding governance substitution in policy-initiated ecosystems. Decoupling refers to the separation between the formal structures organizations display to their institutional environment and the actual activities they produce. Organizations facing strong institutional pressures to adopt specific governance forms often adopt those forms ceremonially, displaying them to external audiences. They also maintain different actual practices in their operational core. For policy-initiated purpose ecosystems, decoupling takes the specific form of purpose-washing. These are formal governance structures that display all the correct markers of purpose-driven, multi-stakeholder governance. The actual decisions are driven by the organizational interests of dominant actors rather than by genuine transformative commitment.

Preventing governance decoupling requires designing formal governance structures that are genuinely aligned with the ecosystem's Transformative Purpose rather than merely compliant with institutional requirements (Battilana & Lee, 2014; Pache & Santos, 2013). This alignment cannot be achieved through governance design alone. It requires the presence of actors who genuinely share the ecosystem's transformative values and are willing to use governance structures as instruments of authentic accountability rather than as shields against external scrutiny. Governance design can, however, create structural conditions that make genuine alignment more likely and governance capture more difficult.

The design of governance bodies should be guided by the principle of stakeholder representation calibrated to purpose relevance rather than to institutional power. Conventional governance board composition for policy-initiated initiatives tends to reflect the power hierarchy of the institutions involved. The largest funders, the most prominent anchor institutions, and the most organizationally sophisticated partners typically receive the most governance seats and influence. For purpose ecosystems, this power-proportionate representation logic systematically underweights the voices of the communities whose transformation the ecosystem is most directly committed to generating. Purpose governance design should actively compensate for this structural bias by reserving governance seats for community representatives, requiring that governance bodies include actors with direct lived experience of the social and environmental challenges the ecosystem addresses, and designing deliberative processes that elevate community knowledge alongside institutional expertise.

The accountability architecture of policy-initiated purpose ecosystems must balance two accountability relationships that can be in tension. On the one hand, upward accountability to funders and institutional overseers, who require demonstrated performance against predetermined objectives. On the other hand, downward accountability to the communities whose transformation the ecosystem is committed to generating, who require genuine responsiveness to their needs, priorities, and assessments of ecosystem performance (Ebrahim & Rangan, 2014; Fung, 2006). Conventional accountability systems prioritize upward accountability. For purpose ecosystems, this prioritization is fundamentally inconsistent with the co-creation logic constitutive of their identity. An ecosystem genuinely accountable to funders but not to communities is not a purpose ecosystem but a funded program using purpose ecosystem language.

Genuine downward accountability requires that communities have real power to shape ecosystem direction, not merely the opportunity to comment on decisions already made. It requires that community assessments of ecosystem performance carry genuine weight in governance decisions, not merely appear in annual reports alongside more authoritative quantitative indicators. And it requires that the ecosystem is genuinely responsive to community feedback. When communities indicate that ecosystem activities are not serving their needs or are generating unintended negative consequences, governance processes respond with genuine adaptation rather than defensive justification.

Purpose tension (the persistent pressure to prioritize the organizational interests of ecosystem participants over the ecosystem's collective Transformative Purpose) is a central governance challenge in policy-initiated ecosystems (Battilana & Dorado, 2010; Pache & Santos, 2013). Universities may be primarily motivated by research funding and publication opportunities. Social enterprises may prioritize financial sustainability over mission delivery. Private firms may participate to access legitimacy and social license. Public agencies may participate to satisfy policy mandates rather than from genuine commitment to the communities the ecosystem serves. Managing purpose tension requires governance mechanisms that maintain the ecosystem's collective purpose as a binding constraint on individual behavior, prevent governance capture by any single actor type, and create accountability for mission delivery, that is, independent of the interests of the most powerful participants.

Specific mechanisms for managing purpose tension include purpose audits (regular, independent assessments of whether ecosystem activities and governance decisions are genuinely aligned with the stated Transformative Purpose, conducted with genuine accountability to community stakeholders), purpose advocates (designated governance roles with explicit responsibility for representing the interests of the ecosystem's intended beneficiaries and raising concerns when organizational interests appear to be overriding purpose commitment), and purpose compacts (explicit agreements among ecosystem participants that specify the behavioral commitments each actor makes in the service of the ecosystem's shared purpose, with clear and credible consequences for non-compliance).

The governance of knowledge and information in policy-initiated ecosystems has direct implications for accountability and purpose integrity. Knowledge asymmetries, where some actors have significantly more information about ecosystem activities, outcomes, and governance decisions than others, create structural conditions for governance capture (Fung et al., 2007; Gaventa, 2006). Actors with superior information can shape governance agendas, frame issues, and anticipate consequences in ways that systematically advantage their interests. Designing governance systems that distribute information equitably among all ecosystem participants, including community members and under-resourced actors, is therefore not merely an accountability requirement but a purpose integrity condition.

Open reporting practices (in which ecosystem performance data and governance decisions are shared publicly and in accessible formats with all stakeholders including community members) create structural conditions for genuine accountability that supplement formal reporting to funders (Fung, 2006; Meijer, 2014). Independent evaluation processes (in which ecosystem performance is assessed by evaluators with genuine accountability to community stakeholders rather than to institutional funders) create accountability relationships that cannot be captured by dominant ecosystem actors. And participatory knowledge co-production processes (in which community members are involved as genuine co-producers of ecosystem knowledge rather than as subjects of institutional data collection) create the informational conditions for genuine downward accountability (Cargo & Mercer, 2008; Cornwall & Jewkes, 1995).

The lifecycle governance of policy-initiated purpose ecosystems requires specific attention to the transition moments at which governance structures must evolve (Ansell & Gash, 2008; Emerson et al., 2012). We refer to the transition from initial institutional setup to genuine multi-stakeholder engagement; the transition from externally funded program to self-sustaining ecosystem; and the transition from the formal governance requirements of the founding policy instrument to the more adaptive and community-responsive governance that genuine purpose alignment requires. The transition to self-sustainability is particularly significant. It forces the ecosystem to develop the resource mobilization capabilities and stakeholder commitment that enable continued operation without external institutional support. Ecosystems that have developed genuine community rootedness and authentic purpose commitment during their funded period typically navigate this transition more successfully than those whose governance has been primarily oriented toward funder accountability (Table 5.2).

Spotlight 5.2 The Preston Model of Community Wealth Building

In 2008, the city of Preston in Lancashire, northwest England, faced a governance crisis. The collapse of a major regeneration initiative, the withdrawal of private investors, and deep public spending cuts had left the city council with halved government grants and no conventional path to economic recovery. What the council did next, in partnership with the Centre for Local Economic Strategies and inspired by the community wealth building work of the Democracy Collaborative in Cleveland, became known as the Preston Model. It is one of the most documented examples of purpose-driven governance transformation in contemporary local government. The model began with a radical governance insight. The city's largest institutions (the university, the hospital, the police, the colleges, the city and county councils) were spending hundreds of millions of pounds annually on procurement, and virtually none of it was circulating within the local economy.

By working collaboratively with these "anchor institutions" to redirect their procurement toward local suppliers, Preston was not creating new money but it was governing the circulation of money that already existed. Between 2012 and 2017, local procurement by the six anchor institutions increased from 39% to 79% of their spending, generating an additional 200 million pounds of economic activity within the Preston economy. The city rose from the bottom quintile of England's most deprived communities to become, by 2016, one of the best places to live and work in the north of England. By 2022, real living wage jobs in Preston had increased from 76% to 88%.

The Preston Cooperative Development Network established worker cooperatives, a Community Land Trust, and initiated work on a regional cooperative bank with a target lending capacity of nearly one billion pounds.

What makes Preston a purpose ecosystem, rather than merely a successful local economic development initiative, is the governance architecture through which the transformation was pursued. The Preston Cooperative Development Network brought together Preston City Council, the University of Central Lancashire, local cooperatives, civic organizations, and community groups in a governance relationship that was genuinely distributed. No single actor controlled the direction of the ecosystem. The council provided convening power and institutional legitimacy; the university provided research capacity; the cooperatives provided democratic governance expertise; the community organizations provided connection to the residents whose economic lives the ecosystem was committed to improving. The accountability architecture is genuinely downward. The ecosystem's governance is oriented toward demonstrating impact on the lives of Preston residents, not toward satisfying the reporting requirements of external funders.

The Preston Model has confronted directly the governance paradox, that is, the tension between formal institutional structures that provide legitimacy and stability, and the authentic community engagement that gives the ecosystem its identity and purpose. As the model has become more recognized and replicated internationally, there is a

Table 5.2 Governance challenges and responses in policy-initiated ecosystems

Governance challenge	Risk if unaddressed	Governance response
Upward vs downward accountability tension	Ecosystem becomes accountable to funders but not to communities	Reserve governance seats for community representatives; require community assessments to carry genuine governance weight
Purpose tension among participants	Organizational interests of dominant actors displace collective purpose	Purpose audits, purpose advocates, and purpose compacts with explicit behavioral commitments
Knowledge asymmetry	Information concentration enables governance capture	Open reporting; participatory knowledge co-production; independent evaluation with community accountability
Governance decoupling	Formal structures perform compliance while activities serve dominant interests	Purpose integrity mechanisms; independent auditing; genuine community accountability with consequential feedback
Purpose drift toward purpose-washing	Gradual erosion through accumulated institutional pressures	Periodic governance reviews with community participation; clear purpose-realignment protocols when drift is detected

Source Authors' own elaboration

genuine risk that its formal structures become ends in themselves rather than instruments of the social transformation that motivated their creation. The council's continuous investment in community participation, its commitment to genuine measurement of impact on inequality and well-being, and its refusal to allow the Preston Model to become a brand rather than a practice, is itself an expression of the governance integrity that this chapter argues is the ultimate test of governance quality in purpose ecosystems.

5.3 Governing for Purpose: Legitimacy, Power, and the Paradox of Structure

Why do actors voluntarily submit to governance without hierarchical authority? How does power operate in systems that claim to distribute it equitably? And what is the fundamental tension between structure and purpose that makes purpose ecosystem governance permanently fragile?

The first question concerns legitimacy. In hierarchical organizations, governance is legitimate because it is authorized. A board of directors has legal authority over management; an elected government has democratic authority over public administration; a manager has contractual authority over employees. In purpose ecosystems, none of these sources of authority is available. No actor has legal, democratic, or contractual authority over others. Yet, actors do submit to governance, do accept constraints on their behavior, and do hold each other accountable for shared commitments.

The management and organizational sociology literatures offer several complementary answers. Suchman's (1995) influential typology of organizational legitimacy distinguishes three sources. First, pragmatic legitimacy, based on the perception that the organization serves the interests of the audience. Then moral legitimacy, related to the perception that the organization is doing the right thing, and cognitive legitimacy based on the assumption that the organization is a natural and necessary feature of the social landscape. For purpose ecosystems, moral legitimacy is the most fundamental and the most distinctive source of governance authority (Selznick, 1957; Tyler, 1990). Actors submit to purpose ecosystem governance not primarily because it serves their interests or because they take it for granted, but because they believe the ecosystem is genuinely committed to values that they share and to social and environmental transformation that they consider genuinely important. Governance authority derives from the authenticity of the shared purpose it serves, not from any formal source of institutional power.

This moral grounding of governance legitimacy has a profound implication. Actually, governance authority in purpose ecosystems is inherently conditional and contestable (Beetham, 1991; Rothstein, 2009). It holds as long as the governance structures are genuinely aligned with the shared purpose that motivates actors to participate and cooperate. When that alignment is questioned, when actors perceive that governance is serving the interests of dominant participants rather than the ecosystem's Transformative Purpose, governance authority erodes. Actors withdraw cooperation, exit the ecosystem, or challenge governance decisions through forms of internal resistance that undermine the ecosystem's collective effectiveness. The most damaging governance crises in purpose ecosystems are not crises of formal authority but crises of legitimacy. We refer to moments when a sufficient number of actors conclude that the governance of the ecosystem is no longer genuinely in the service of the purpose that justified their participation.

Habermas's (1984) theory of communicative action provides the deepest philosophical foundation for understanding governance legitimacy in purpose ecosystems. Habermas distinguishes between strategic action (oriented toward the achievement of individual goals) and communicative action (oriented toward achieving genuine mutual understanding through reasoned discourse). Governance in purpose ecosystems, at its best, is an instance of communicative action where actors engage with each other not to pursue their individual interests through strategic manipulation but to develop genuine shared understanding of what their collective purpose requires and how it can best be served. This communicative foundation of governance legitimacy explains its strength and its fragility. Its strength lies in the genuine commitment it generates when it works. Its fragility lies in its dependence on the continued willingness of actors to engage in genuine discourse rather than strategic manipulation.

The second question concerns power. Purpose ecosystems claim to distribute power equitably among diverse actors, to give voice to marginalized communities, and to prevent any single actor from dominating governance direction. Purpose ecosystems that are captured by the interests of their most powerful participants are not purpose ecosystems in any meaningful sense, regardless of their governance documents (Clegg, 1989; Lukes, 2005). But these claims are also frequently idealized. In practice, power in purpose ecosystems is distributed unevenly, and the gap between the formal equality of governance participation and the substantive inequality of governance influence is one of the most persistent and most consequential features of multi-actor collaborative governance.

Gaventa's (2006) power cube framework analyzes power along three dimensions. These are the visible power, that represents the formal rules and decision-making, the hidden power, that is the agenda-setting and issue framing, and the invisible power, so norms and beliefs that shape what is considered possible and legitimate. Visible power is most easily addressed through formal governance design, such as reserved seats for community representatives, voting rights for marginalized groups, transparent decision-making processes. But it is also the least consequential form of power in practice. Actors who control hidden power (who determine which issues appear on governance agendas and how they are framed) exercise far more influence over ecosystem direction than formal voting rights reveal. And actors who exercise invisible power (who shape the norms and beliefs that determine what governance participants perceive as possible and legitimate) exercise the most profound and most durable form of governance influence. Addressing all three levels is the real challenge of equitable governance design.

Addressing hidden and invisible power requires governance practices that go beyond formal representation mechanisms (Fung, 2006; Innes & Booher, 2010). Deliberative processes that give community members genuine influence over agenda-setting (not merely over agenda items once they have been framed by institutional actors) address hidden power. Governance cultures that actively challenge the dominant assumptions about whose knowledge counts as expertise, whose priorities count as legitimate governance concerns, and whose visions of transformation count as realistic aspirations address invisible power (Collins, 1990; Freire, 1970). And longitudinal investment in the organizational capacity of under-resourced governance participants (enabling them to engage with the analytical resources and institutional confidence that better-resourced actors take for granted) addresses the structural roots of governance power inequality.

Young's (1990) concept of the five faces of oppression (exploitation, marginalization, powerlessness, cultural imperialism, and violence) provides a comprehensive taxonomy of the forms of structural injustice that purpose ecosystem governance must actively resist rather than passively reproduce. Cultural imperialism is particularly relevant for purpose ecosystem governance. The imposition of dominant cultural frameworks, professional languages, and institutional logics on governance processes that are nominally inclusive systematically disadvantages governance participants from different cultural backgrounds, community traditions, and ways of knowing. Governance, that is, culturally imperialist in practice, even when formally inclusive in design, will systematically exclude the perspectives and priorities of the communities whose transformation the ecosystem is most fundamentally committed to generating.

The third question concerns the paradox of structure. Structure (governance rules, accountability mechanisms, formal procedures, institutional arrangements) is what makes purpose collective action possible at scale (North, 1990; Ostrom, 1990). Without it, multi-actor collaboration is vulnerable to free-riding, opportunistic behavior, and the gradual erosion of shared commitments under the pressure of individual interests. Structure creates predictability, accountability, and the institutional memory that enables collective action to persist through the inevitable turnover of individual actors. In this sense, structure is the friend of purpose. It protects the ecosystem's transformative commitment from the entropy that threatens all voluntary collective enterprises.

But structure is also the enemy of purpose in a profound and irreducible sense. Every governance structure creates incentives for compliance performance rather than genuine engagement (Espeland & Sauder, 2007; Power, 1999). Every formal accountability mechanism creates incentives to optimize for measured indicators rather than for unmeasured but more important dimensions of transformative impact. Every institutional arrangement creates organizational interests in its own survival and reproduction, which over time can diverge from the Transformative Purposes the arrangement was created to serve (Michels, 1911; Selznick, 1957). The more elaborate and codified the governance structure, the more powerful these distorting incentives become.

DiMaggio and Powell's (1983) analysis of institutional isomorphism explains how this happens. Isomorphic pressures (coercive pressures from funders and regulators, mimetic pressures to adopt the governance forms of successful comparable organizations, and normative pressures from professional communities that define best governance practice) drive purpose ecosystems toward governance structures that are valued by their institutional environment rather than by the communities whose transformation they are committed to generating. The result is governance that looks increasingly professional, increasingly formalized, and increasingly similar to the governance of conventional institutional actors. This governance becomes decreasingly responsive to the informal, relational, and community-rooted values that gave the ecosystem its distinctive identity and transformative potential.

The paradox of structure has no final resolution. It is a permanent tension that every purpose ecosystem must manage rather than solve (March & Olsen, 1989; Weick, 1979). Managing it requires what we might call structural reflexivity. This is the continuous practice of questioning whether the governance structures of the ecosystem are serving or distorting its Transformative Purpose, and revising those structures when the answer is the latter. Structural reflexivity is not comfortable. It requires governance actors to challenge arrangements that benefit them, to acknowledge when formal accountability is substituting for genuine purpose engagement, and to accept the uncertainty and instability that comes with governance structures that are genuinely responsive to evolving community needs rather than to the procedural requirements of institutional legitimacy.

Selznick's (1957) concept of institutional character provides a valuable complement to structural reflexivity as a governance principle. Selznick argued that organizations develop distinctive institutional characters (sets of values and commitments infused throughout the organization's identity that shape the behavior of organizational members even in the absence of explicit instructions). For purpose ecosystems, developing a strong institutional character oriented toward genuine transformative commitment is the most durable protection against the distorting effects of governance structure. Ecosystems whose governance culture genuinely embodies the values of purpose, community accountability, equity, and adaptive learning are more resistant to isomorphic pressures than those whose governance culture is primarily oriented toward compliance and institutional legitimacy. Building institutional character takes time and cannot be accomplished through governance design alone. It requires sustained investment in the relational, deliberative, and community engagement practices through which genuine purpose commitment is continuously renewed and deepened.

The four governance functions introduced in Chap. 1 (purpose recognition, participation design, accountability, and adaptive governance) can now be understood in the light of the theoretical analysis of this section. Purpose recognition is the governance function through which the legitimacy-grounding commitment to shared purpose is continuously renewed. It creates the deliberative spaces where actors can reconnect with the values that motivate their participation and reaffirm the collective commitments that give governance its authority. Participation design is the governance function through which power asymmetries are actively addressed. It creates the structural conditions that enable marginalized voices to shape governance direction rather than merely ratify decisions made by dominant actors. Accountability maintains the alignment between governance structure and transformative commitment. And adaptive governance is the function through which the paradox of structure is managed. It ensures that governance structures remain responsive to evidence and evolving community needs rather than becoming entrenched ends in themselves.

Together, these four functions constitute what might be called purpose governance. This is a governance genuinely in the service of the ecosystem's transformative commitment, distributing power with honest attention to its structural dimensions, maintaining its legitimacy through genuine alignment with shared values rather than through formal authority, and managing the paradox of structure through continuous structural reflexivity. Purpose governance is not a state that ecosystems achieve and maintain. It is a practice. One that must be continuously renewed through the deliberate choices of the actors who inhabit governance structures and the communities whose lives those structures are meant to serve.

Box 5.1 The Three Foundational Questions of Purpose Ecosystem Governance

Governance authority in purpose ecosystems is derived from the authenticity of the shared purpose it serves, not from formal institutional power. It is moral legitimacy, not legal or democratic authority, that induces voluntary compliance. This makes governance authority stronger and more fragile than conventional authority; stronger because it generates genuine commitment; more fragile because it is conditional on continued alignment between governance structures and the purposes that justify them.

Power in purpose ecosystems operates at three levels, the visible power related to formal decision-making, the hidden power related to agenda-setting and framing, and invisible power related to norms and beliefs about what is possible and legitimate. Formal governance design addresses only visible power. Genuine equity in governance requires deliberate attention to all three levels, and sustained investment in the organizational capacity of under-resourced participants to engage in governance processes on genuinely equal terms.

Structure protects purpose by creating predictability, accountability, and institutional memory. But it also distorts purpose by creating incentives for compliance performance rather than genuine engagement, and by generating isomorphic pressures toward governance forms valued by institutional environments rather than by transforming communities. The paradox has no final resolution. It must be managed through structural reflexivity, the continuous practice of questioning whether governance structures are serving or distorting the Transformative Purpose they are meant to protect.

References

Adner, R. (2017). Ecosystem as structure: An actionable construct for strategy. *Journal of Management, 43*(1), 39–58.

Ahuja, G. (2000). Collaboration networks, structural holes, and innovation: A longitudinal study. *Administrative Science Quarterly, 45*(3), 425–455.

Ansell, C., & Gash, A. (2008). Collaborative governance in theory and practice. *Journal of Public Administration Research and Theory, 18*(4), 543–571.

Argyris, C., & Schön, D. A. (1978). *Organizational learning: A theory of action perspective*. Addison-Wesley.

Autio, E., & Thomas, L. D. W. (2019). Innovation ecosystems. In R. Adner, J. E. Oxley, & B. S. Silverman (Eds.), *Advances in strategic management: Collaboration and competition in business ecosystems*. Emerald Publishing.

Battilana, J., & Dorado, S. (2010). Building sustainable hybrid organizations: The case of commercial microfinance organizations. *Academy of Management Journal, 53*(6), 1419–1440.

Battilana, J., & Lee, M. (2014). Advancing research on hybrid organizing: Insights from the study of social enterprises. *Academy of Management Annals, 8*(1), 397–441.

Beetham, D. (1991). *The legitimation of power*. Macmillan.

Bourdieu, P. (1977). *Outline of a theory of practice*. Cambridge University Press.

Cargo, M., & Mercer, S. L. (2008). The value and challenges of participatory research: Strengthening its practice. *Annual Review of Public Health, 29*, 325–350.

Clegg, S. R. (1989). *Frameworks of power*. SAGE.

Collins, P. H. (1990). *Black feminist thought: Knowledge, consciousness, and the politics of empowerment*. Routledge.

Cornwall, A., & Jewkes, R. (1995). What is participatory research? *Social Science & Medicine, 41*(12), 1667–1676.

DiMaggio, P. J., & Powell, W. W. (1983). The iron cage revisited: Institutional isomorphism and collective rationality in organizational fields. *American Sociological Review, 48*(2), 147–160.

Ebrahim, A., & Rangan, V. K. (2014). What impact? A framework for measuring the scale and scope of social performance. *California Management Review, 56*(3), 118–141.

Emerson, K., Nabatchi, T., & Balogh, S. (2012). An integrative framework for collaborative governance. *Journal of Public Administration Research and Theory, 22*(1), 1–29.

Espeland, W. N., & Sauder, M. (2007). Rankings and reactivity: How public measures recreate social worlds. *American Journal of Sociology, 113*(1), 1–40.

Freire, P. (1970). *Pedagogy of the oppressed*. Herder and Herder.

Fung, A. (2006). Varieties of participation in complex governance. *Public Administration Review, 66*(s1), 66–75.

Fung, A., Graham, M., & Weil, D. (2007). *Full disclosure: The perils and promise of transparency*. Cambridge University Press.

Gaventa, J. (2006). Finding the spaces for change: A power analysis. *IDS Bulletin, 37*(6), 23–33.

Gawer, A., & Cusumano, M. A. (2014). Industry platforms and ecosystem innovation. *Journal of Product Innovation Management, 31*(3), 417–433.

Giddens, A. (1984). *The constitution of society: Outline of the theory of structuration*. University of California Press.

Granovetter, M. (1985). Economic action and social structure: The problem of embeddedness. *American Journal of Sociology, 91*(3), 481–510.

Gray, B. (1989). *Collaborating: Finding common ground for multiparty problems*. Jossey-Bass.

Habermas, J. (1984). *The theory of communicative action*. Beacon Press.

Huxham, C., & Vangen, S. (2005). *Managing to collaborate: The theory and practice of collaborative advantage*. Routledge.

Innes, J. E., & Booher, D. E. (2010). *Planning with complexity: An introduction to collaborative rationality for public policy*. Routledge.

Jacobides, M. G., Cennamo, C., & Gawer, A. (2018). Towards a theory of ecosystems. *Strategic Management Journal, 39*(8), 2255–2276.
Kania, J., & Kramer, M. (2011). Collective impact. *Stanford Social Innovation Review, 9*(1), 36–41.
Lave, J., & Wenger, E. (1991). *Situated learning: Legitimate peripheral participation.* Cambridge University Press.
Luhmann, N. (1979). *Trust and power.* John Wiley and Sons.
Lukes, S. (2005). *Power: A radical view* (2nd ed.). Palgrave Macmillan.
March, J. G., & Olsen, J. P. (1989). *Rediscovering institutions: The organizational basis of politics.* Free Press.
Meijer, A. J. (2014). Transparency. In M. Bovens, R. E. Goodin, & T. Schillemans (Eds.), *The Oxford handbook of public accountability* (pp. 507–524). Oxford University Press.
Meyer, J. W., & Rowan, B. (1977). Institutionalized organizations: Formal structure as myth and ceremony. *American Journal of Sociology, 83*(2), 340–363.
Michels, R. (1911). *Political parties: A sociological study of the oligarchical tendencies of modern democracy.* Jarrold & Sons.
North, D. C. (1990). *Institutions, institutional change and economic performance.* Cambridge University Press.
Ostrom, E. (1990). *Governing the commons: The evolution of institutions for collective action.* Cambridge University Press.
Ostrom, E. (2010). Beyond markets and states: Polycentric governance of complex economic systems. *American Economic Review, 100*(3), 641–672.
Pache, A. C., & Santos, F. (2013). Inside the hybrid organization: Selective coupling as a response to competing institutional logics. *Academy of Management Journal, 56*(4), 972–1001.
Parker, G. G., Van Alstyne, M. W., & Choudary, S. P. (2016). *Platform revolution: How networked markets are transforming the economy and how to make them work for you.* Norton.
Power, M. (1999). *The audit society: Rituals of verification.* Oxford University Press.
Provan, K. G., & Kenis, P. (2008). Modes of network governance: Structure, management, and effectiveness. *Journal of Public Administration Research and Theory, 18*(2), 229–252.
Rothstein, B. (2009). Creating political legitimacy: Electoral democracy versus quality of government. *American Behavioral Scientist, 53*(3), 311–330.
Schatzki, T. R., Knorr-Cetina, K., & von Savigny, E. (Eds.). (2001). The practice turn in contemporary theory. Routledge.
Selznick, P. (1957). *Leadership in administration.* Harper and Row.
Skelcher, C., & Smith, S. R. (2015). Theorizing hybridity: Institutional logics, complex organizations, and actor identities. *Public Administration, 93*(2), 433–448.
Spigel, B. (2017). The relational organization of entrepreneurial ecosystems. *Entrepreneurship Theory and Practice, 41*(1), 49–72.
Stam, E. (2015). Entrepreneurial ecosystems and regional policy: A sympathetic critique. *European Planning Studies, 23*(9), 1759–1769.
Suchman, M. C. (1995). Managing legitimacy: Strategic and institutional approaches. *Academy of Management Review, 20*(3), 571–610.
Susskind, L., & Cruikshank, J. (1987). *Breaking the impasse: Consensual approaches to resolving public disputes.* Basic Books.
Tyler, T. R. (1990). *Why people obey the law.* Yale University Press.
Weick, K. E. (1979). *The social psychology of organizing* (2nd ed.). McGraw-Hill.
Wenger, E. (1998). *Communities of practice: Learning, meaning, and identity.* Cambridge University Press.
Young, I. M. (1990). *Justice and the politics of difference.* Princeton University Press.

Chapter 6
Measuring Impact in Purpose Ecosystems

The question whether purpose ecosystem direction is generating genuine, durable, and system-level change in the social and environmental conditions of a territory refers to impact. Impact measurement is not the same as the output and short-term outcome monitoring. Output and short-term outcome tell us whether an ecosystem is moving in the right direction, namely the programs delivered, the partnerships formed, the early changes in conditions and behaviors that suggest the ecosystem is working. Impact refers to the long-term, systemic, and often difficult-to-attribute changes in social, environmental, territorial, and institutional conditions that represent the ultimate test of whether a purpose ecosystem has genuinely realized its Transformative Purpose.

This distinction reflects a fundamental theoretical commitment about the nature of purpose ecosystem impact. Impact here is not the property of any single actor. It is not what a university contributes, what a social enterprise delivers, or what a public agency funds. It is what the ecosystem as a whole generates through its collective activities, relationships, and institutional influence over time. It is emergent because it arises from the interactions of many actors, accumulates gradually through processes that are often invisible in the short term, and belongs to the community and territory rather than to any of the organizations that contributed to it. Understanding this collective, emergent, and territorial character of ecosystem impact is the theoretical foundation on which the entire measurement framework of this chapter rests.

The challenge of measuring impact in multi-actor systems has attracted growing attention in recent years. Ebrahim and Rangan (2014) demonstrate that not all organizations should aspire to measure long-term systemic impact, since the causal relation between their activities and the systemic change they produce is too long, complex, and context-dependent to claim credible attribution. This means that the appropriate unit of impact assessment could be the ecosystem and its territory, not any individual actor within it. A second strand concerns the relationship between measurement and governance. Ebrahim (2019) argues that measurement systems in complex organizations inevitably shape what actors attend to, what they prioritize. Power (1999)

M. Nicotra and M. Romano, *Purpose Ecosystems*, Contributions to Management Science, https://doi.org/10.1007/978-3-032-38382-2_6

describes this as the performative character of audit, and this could also be applied to ecosystems at the expense of their genuine Transformative Purpose. A third strand explicitly addresses the measurement of systemic and ecosystem-level change. Krlev et al. (2020) and Pel et al. (2020) argue that transformative social innovation requires measurement frameworks capable of capturing not only direct outputs and outcomes but the deeper shifts in norms, institutional practices, and power relations that constitute genuine systems change. These contributions collectively reinforce the theoretical architecture of this chapter that is organized into three sections. Section 6.1 develops the conceptual foundations for understanding ecosystem impact, distinguishing it from output monitoring and individual actor performance, explaining its emergent and systemic character, and introducing the distinction between the observational measurement that characterizes organically emergent ecosystems and the formal measurement that characterizes policy-initiated ones. Section 6.2 presents a unified framework for ecosystem impact assessment organized around four long-term impact dimensions: environmental sustainability transitions, social transformation, territorial development, and ecosystem institutional legacy. Section 6.3 addresses the practical dimensions of ecosystem impact assessment that is how to handle the attribution challenge in complex systems, how impact assessment can feed governance and collective learning, and how honest measurement builds the legitimacy that purpose ecosystems depend on.

6.1 The Nature of Ecosystem Impact: Emergence, System-Level Change, and Long-Term Transformation

Impact has a long history as a concept in the fields of organizational management, public policy, evaluation research, and social innovation. In its most general sense, it refers to the changes in conditions, systems, or well-being that can be attributed to a specific intervention, program, or organization. This general definition conceals a range of different understandings of what change means, at what level it should be assessed, over what time horizon, and through what causal mechanisms it is generated. These differences matter enormously for purpose ecosystems, because the conventional impact assessment frameworks developed for organizations, projects, and programs are systematically inadequate for assessing the kind of change that purpose ecosystems exist to generate.

The most important inadequacy of conventional impact assessment frameworks, when applied to purpose ecosystems, is their assumption that impact is a property of individual actors or interventions. Social Return on Investment (SROI) calculates the social value generated by a specific organization or program. Theory of Change frameworks trace the causal logic from the activities of a specific intervention to its intended outcomes. ESG frameworks measure the environmental, social, and governance performance of individual companies. Impact investing frameworks assess the social and environmental outcomes of specific investment portfolios. All of these approaches share the common assumption that impact can be attributed to a specific actor, that the relevant unit of analysis is the organization or program, and that measurement means tracing the contribution of a specific intervention to observed changes in conditions.

In purpose ecosystems, there is not a single actor that can claim the impact of a purpose ecosystem as its own. In the same way, impact is not reducible to the sum of these individual contributions. The university that trains skilled workers, the social enterprise that employs them, the civic organization that advocates for policy change, the public institution that funds infrastructure, and the community members who participate in governance and co-creation all contribute to the ecosystem's impact. In this context, the development of composite indicator approaches capable of capturing these multi-dimensional elements is relevant and represents a frontier of ecosystem research (Corrente et al., 2019; Nicotra, 2019; Romano et al., 2020, 2025).

This emergent character of ecosystem impact has direct implications for measurement. It means that conventional attribution methodologies, which try to connect specific outputs to specific outcomes through linear causal chains, are poorly suited to ecosystem contexts. It means that the relevant unit of analysis is the ecosystem and its territory, not any individual organization. Moreover, the relevant time horizon is years to decades, not months or project cycles. And it means the most important dimensions of impact are the most resistant to measurement through conventional quantitative indicators.

The distinction between output monitoring, short-term outcome tracking, and long-term impact assessment is fundamental to the architecture of this book. Output indicators capture the immediate, directly attributable products of ecosystem activities, namely the programs delivered, the partnerships formed, the hectares protected, the services provided. Short-term outcome indicators capture the first signals of change in conditions, behaviors, and capabilities. These include improvements in skills, reductions in exclusion, early changes in environmental performance. These indicators are valuable for navigation. They tell us whether the ecosystem is active and whether its activities are beginning to generate the expected early results.

Long-term impact indicators, as developed in this chapter, operate at a fundamentally different level. They capture systemic changes in the social, environmental, territorial, and institutional conditions of a territory that emerge from the cumulative effect of ecosystem activities over years to decades. They include things like permanent reductions in territorial inequality, transitions to low-carbon energy systems, transformations in the governance norms and institutional practices of a territory, and the development of new forms of social capital and civic capacity that outlast any specific ecosystem initiative. These changes are not directly attributable to specific ecosystem activities in the way that outputs are. They are the product of complex causal processes in which ecosystem activities interact with many other factors (including policy changes, economic shifts, technological developments, and cultural evolution) to produce systemic change. Demonstrating that a purpose ecosystem has contributed to these changes requires different methodological approaches than those appropriate for output monitoring.

Understanding who measures ecosystem impact, and why, requires distinguishing between two fundamentally different contexts: organically emergent ecosystems and policy-initiated ecosystems.

In organically emergent ecosystems (those that arise from the bottom-up interaction of actors who share values and a vision of change without being formally constituted by any policy instrument) impact measurement is not a function of internal governance. There is no central structure with the authority and the capacity to produce comprehensive impact assessments. Impact, in these ecosystems, is observed from the outside by actors with different interests and different capacities for observation. Communities observe impact through the changes they experience in their territory, such as cleaner air, better employment opportunities, more vibrant civic life, stronger social networks, greater sense of agency over their collective future. They do not produce formal impact reports, but they develop collective narratives of change that are themselves a form of impact assessment, however informal. Policymakers observe impact through the lens of their territorial development responsibilities. Researchers develop theories about how ecosystems generate impact, test them against empirical evidence, and contribute to the cumulative knowledge base that helps future ecosystems be more effective. Potential new actors assess whether an ecosystem is generating the kind of change they want to contribute to. Investors and funders want to know whether their investments are generating the social and environmental returns they seek.

What is common to all these forms of observational impact assessment in emergent ecosystems is that they are motivated by orientation and learning rather than by control. No external observer has the authority to direct the ecosystem based on impact evidence, because there is no authority in an emergent ecosystem to direct. Impact evidence in emergent ecosystems serves to inform decisions about whether to engage with, support, or build on the ecosystem, not to control its activities. This is a fundamentally different relationship between measurement and governance than that which characterizes conventional organizational performance management.

In policy-initiated ecosystems (those created and supported by formal policy instruments such as the Italian PNRR national ecosystems, the European Extended Partnerships, or the territorial development programs associated with Smart Specialization Strategies) the situation is fundamentally different. These ecosystems have formal governance structures, contractual accountability relationships, and reporting obligations that create the capacity and the necessity for systematic impact measurement. The institutions that fund and oversee these ecosystems have a legitimate interest in knowing whether public investments are generating the expected social and environmental returns, and the formal governance structures of the ecosystems have the mandate and the organizational capacity to collect, analyze, and report impact data systematically. In policy-initiated ecosystems, impact measurement serves multiple functions simultaneously. It provides accountability to funders and the public, it informs governance decisions about resource allocation and strategic direction, it generates learning that can improve ecosystem effectiveness over time, and it contributes to the evidence base that informs future policy design.

The distinction between observational measurement in emergent ecosystems and formal measurement in policy-initiated ones does not imply that the relevant dimensions of impact are different. Both types of ecosystems aspire to generate the same kinds of long-term systemic change, that is, environmental sustainability transitions, social transformation, territorial development, and institutional legacy. The difference lies in who measures, how systematically, and for what purpose. The unified impact assessment framework developed in Sect. 6.2 is designed to be applicable in both contexts, with the understanding that the rigor, systematicity, and formality of its implementation will vary depending on whether the ecosystem has the governance structures needed for formal measurement or must rely on more informal and distributed forms of observational assessment.

Before presenting the framework, it is important to introduce the methodological foundations that make impact assessment in complex, multi-actor systems possible and credible. Three traditions are particularly relevant.

Complexity theory and evaluation, developed by scholars such as Patton (2011), Rogers (2008), and Ramalingam et al. (2008), provides the conceptual foundation for understanding why conventional linear evaluation approaches are inadequate for purpose ecosystems and what alternatives are available. Complex systems are characterized by non-linearity, emergence, feedback loops, and context-dependence. The same intervention can produce very different outcomes in different contexts, and outcomes often emerge from the interaction of multiple factors in ways that cannot be predicted from the properties of any single component. Evaluation approaches designed for linear causal chains (where specific inputs produce specific outputs through predictable mechanisms) systematically misrepresent the causal processes through which complex systems generate impact. Complexity-appropriate evaluation approaches embrace uncertainty, focus on patterns of change rather than pre-specified outcomes, use multiple methods to triangulate findings, and build in mechanisms for continuous learning and revision.

Contribution analysis, developed by Mayne (2001) and refined by Befani and Mayne (2014), provides the most rigorously developed methodology for demonstrating that a specific intervention or organization has contributed to observed outcomes in complex, multi-causal settings. Contribution analysis does not attempt to prove that a specific actor caused an observed change, which is rarely possible in complex systems. Instead, it aims to demonstrate that the actor made a meaningful contribution to that change, through a combination of theory-based reasoning about the causal mechanisms involved and empirical evidence about how those mechanisms operated in practice. The methodology involves developing a contribution story (a narrative account of how ecosystem activities are expected to contribute to observed changes through specific causal mechanisms) and then testing that story against empirical evidence about whether the mechanisms operated as expected, whether alternative explanations can be ruled out, and whether the evidence of contribution is credible. For purpose ecosystems, contribution analysis is particularly appropriate because it acknowledges the multi-causal complexity of ecosystem impact. At the same time, it provides a rigorous basis for making meaningful claims about ecosystem contribution to observed systemic changes.

Longitudinal and comparative methods provide the empirical infrastructure for long-term impact assessment in purpose ecosystems. Longitudinal studies track changes in ecosystem and territorial conditions over extended time periods, enabling the detection of cumulative effects that are invisible in cross-sectional snapshots. Comparative case studies, which systematically compare ecosystems with different characteristics, different territorial contexts, and different impact trajectories, enable more robust inferences about the conditions under which specific impact dimensions are generated (Eisenhardt & Graebner, 2007; Yin, 2014). Quasi-experimental methods, including difference-in-differences analysis, synthetic control methods, and regression discontinuity designs, can provide credible estimates of ecosystem impact in policy-initiated contexts where data on comparison territories are available (Morgan & Winship, 2015). Participatory evaluation approaches (which involve community members as active co-producers of impact knowledge rather than passive subjects of external assessment) generate evidence, that is, both more relevant to community needs and more legitimate in the eyes of the communities whose transformation is being assessed (Fals Borda, 1987; Reason & Bradbury, 2008).

Box 6.1 Three Levels of Ecosystem Assessment

Output indicators: Immediate, directly attributable products of ecosystem activities. Observable in the short term. Produced by ecosystem actors themselves or observed by external monitors (e.g., number of programs delivered, partnerships formed, hectares protected, services provided).

Short-term outcome indicators: First changes in conditions, behaviors, and capabilities resulting from ecosystem activities. Observable within months to a few years. Require survey and behavioral observation methods (e.g., skill improvements, early reductions in emissions, initial changes in civic participation).

Long-term impact indicators: Systemic, structural, and often emergent changes in social, environmental, territorial, and institutional conditions. Observable over years to decades. Require longitudinal, comparative, and contribution analysis methods. Belong to the community and territory rather than to individual actors (e.g., permanent reductions in inequality, completed energy transitions, transformation of governance norms, development of new institutional forms).

Spotlight 6.1 Ashoka and the Challenge of Systemic Impact Assessment

Ashoka was founded in 1980 by Bill Drayton in Washington DC with a purpose that was simple to state and genuinely difficult to operationalize: to build a world where everyone is a changemaker. Drayton's insight was that the fundamental challenge of social transformation is not the absence of solutions but the absence of the organizational infrastructure, the relational networks, and the cultural norms that enable good solutions to spread, adapt, and generate systemic change. Ashoka's mission, from the beginning, was not to solve specific social problems but to build the ecosystem of social innovation: to identify, support, and connect social entrepreneurs whose work contributes to the broader transformation of social systems. Today Ashoka operates in over 90 countries, supporting more than 3900 elected Fellows and building networks that reach millions of changemakers.

Ashoka's approach to impact measurement reflects this systemic orientation in ways directly relevant to the framework developed in this chapter. The organization makes an explicit and carefully maintained distinction between three levels of impact: individual change (the direct beneficiaries of a social entrepreneur's work), institutional change (the transformation of policies, organizational practices, and sector-level norms), and systems change (the fundamental shifts in cultural assumptions, power relations, and structural conditions that determine what is possible for communities over the long term). Ashoka does not count beneficiaries as its primary impact metric. It tracks the extent to which the social entrepreneurs it supports are generating institutional and systems change, such as changing laws, shifting sector practices, building new field-level infrastructure, and contributing to the cultural evolution of what societies believe is possible.

The methodology Ashoka has developed for assessing systems change impact directly addresses the attribution problem that this chapter identifies as the central challenge of long-term ecosystem impact assessment. Ashoka does not claim that its social entrepreneurs cause systems change in any simple causal sense. It traces contribution pathways, the specific mechanisms through which a social entrepreneur's work has influenced the thinking and practice of other actors, changed institutional arrangements, created new organizational forms, or shifted cultural norms. These contribution pathways are documented through longitudinal case studies, network analysis of influence relationships, and comparative assessment of what institutional arrangements looked like before and after the social entrepreneur's intervention.

The relational dimension of Ashoka's impact framework is particularly relevant for purpose ecosystem impact assessment. Ashoka explicitly treats the quality of the network it has built (measured not in numbers of Fellows but in the density, diversity, and purpose direction of collaborative relationships among them) as a primary impact indicator. The lesson Ashoka offers is that the most important impact indicators are often the hardest to measure: the quality of collaborative relationships, the extent to which shared purpose is genuinely constitutive of individual actors' identities, the degree to which the ecosystem has generated new institutional forms that outlast individual actors, and the cultural shifts that have made previously impossible social arrangements thinkable and achievable. None of these dimensions is reducible to quantitative indicators, and all require longitudinal, qualitative, and participatory assessment methods that most impact measurement frameworks do not accommodate.

6.2 A Unified Framework for Ecosystem Impact Assessment

The unified framework for ecosystem impact assessment is designed to capture the long-term, systemic changes that purpose ecosystems aspire to generate across four dimensions: environmental sustainability transitions, social transformation, territorial development, and ecosystem institutional legacy. These four dimensions correspond to and extend the four dimensions introduced for short-term assessment, but their content changes fundamentally when considered over the long-time horizons relevant for impact rather than output measurement. Empirical approaches that combine quantitative indicators with stochastic multi-attribute analysis have demonstrated the feasibility of assessing ecosystem performance across multiple dimensions simultaneously (Corrente et al., 2019; Romano et al., 2020, 2025).

The framework is explicitly unified. It applies to organically emergent ecosystems and policy-initiated ones. The difference between the two contexts lies not in the dimensions of impact that matter but in how evidence about those dimensions is collected, by whom, and for what purpose. In emergent ecosystems, the framework serves primarily as a conceptual map that helps diverse observers (communities, researchers, policymakers, and potential new actors) make sense of the changes they observe in a territory and assess whether those changes are consistent with the ecosystem's Transformative Purpose. In policy-initiated ecosystems, the framework serves as the basis for a formal monitoring and evaluation system that produces systematic evidence for accountability, governance, and learning.

For each of the four impact dimensions, the framework identifies the nature of the long-term change being assessed, the theoretical frameworks that provide the evaluative criteria, the specific indicators that can be used to assess progress, the data sources and methods most appropriate for each indicator, and the time horizons over which change can be expected to become visible. The indicators are organized into two tiers: a core set of five to seven indicators that every ecosystem should consider for each dimension, regardless of its specific context, and a set of context-specific supplementary indicators that may be relevant for ecosystems operating in particular territorial or sectoral contexts.

Before presenting the four dimensions, it is important to clarify the distinction between contribution and causation. None of the indicators presented below can be attributed exclusively to ecosystem activities. Long-term systemic changes in social, environmental, territorial, and institutional conditions are always the product of multiple causal factors, including policy changes, economic shifts, demographic trends, technological developments, and cultural evolution. What purpose ecosystems can claim (and what the indicators below are designed to assess) is a meaningful and identifiable contribution to those changes. The question the framework asks is not "did the ecosystem cause this change?" but "did the ecosystem make a significant contribution to this change, through identifiable mechanisms, in ways that would not have occurred, or would have occurred more slowly or less completely, without the ecosystem's activities?" This is the contribution logic that underpins the entire framework.

6.2.1 *Environmental Sustainability Transitions*

The long-term environmental impact of a purpose ecosystem is assessed not at the level of individual organizational practices (which are captured by the output and short-term outcome indicators of Chap. 3) but at the level of systemic transitions in the environmental configuration of a territory. A purpose ecosystem that has genuinely contributed to long-term environmental impact will have helped to change the energy system of its territory, to embed circular economy principles in the production and consumption patterns of its economic actors, to restore and protect the ecological assets of its landscape, and to build the institutional and cultural foundations for sustained environmental stewardship. These are transitions that take years to decades to realize. They cannot be produced by any single organization, however committed. They are precisely the kind of systemic change that purpose ecosystems, with their multi-actor composition and territorial embeddedness, are uniquely positioned to advance.

The multi-level perspective on sustainability transitions, developed by Geels (2002) and discussed in Chap. 3, provides the most appropriate theoretical framework for understanding what long-term environmental impact means at the ecosystem level. A regime transition occurs when niche innovations break through to reconfigure the established sociotechnical regime that organizes a sector or territory's energy, food, mobility, or production system. For purpose ecosystems, this means that long-term environmental impact is assessed not by measuring the individual environmental performance of ecosystem organizations but by asking whether the ecosystem has contributed to regime-level changes in the territorial systems within which it operates. Has the energy mix of the territory shifted significantly toward renewables? Have circular economy practices become standard in the sector or territory? Have land use patterns changed in ways that protect and restore biodiversity? Have environmental governance norms and institutional practices changed in ways that make future environmental progress more likely?

Planetary boundaries, developed by Rockstrom et al. (2009), provide a second evaluative reference for long-term environmental impact assessment. Progress toward operating within planetary boundaries provides an ultimate criterion of environmental impact that transcends the measurement of specific organizational activities and connects ecosystem impact to the global environmental challenge. For purpose ecosystems operating in territories with specific environmental challenges, the relevant planetary boundaries (whether climate, biodiversity, land use, water, or biogeochemical cycles) provide the reference framework for assessing whether territorial environmental transitions are moving in the direction of global sustainability.

Natural capital accounting, developed by scholars such as Costanza et al. (1997, 2014) and operationalized through frameworks such as the System of Environmental-Economic Accounting (SEEA) developed by the United Nations, provides a methodology for quantifying the stock and flow of natural capital in a territory. Long-term improvements in natural capital stocks (including forest cover, soil quality, water availability, biodiversity, and coastal ecosystem health) provide direct evidence of ecosystem-level environmental impact that goes beyond the measurement of pollution reduction or energy efficiency. Purpose ecosystems operating in territories with significant natural capital assets can use natural capital accounting to demonstrate their contribution to the preservation and enhancement of those assets over time.

Table 6.1 summarizes long-term impact indicators related to environmental sustainability transitions.

Table 6.1 Long-term impact indicators: environmental sustainability transitions

Long-term impact indicator	What it measures	Time horizon	Primary method
Territorial energy transition progress	Share of renewable energy in territorial energy mix; rate of fossil fuel displacement; investment flows toward low-carbon infrastructure	5–15 years	Energy statistics; investment tracking
Scope 3 territorial emission trajectory	Direction and rate of change in territorial GHG emissions including supply chain effects; alignment with science-based targets	5–20 years	Territorial GHG inventories; LCA studies
Circular economy regime penetration	Share of territorial economic activity organized according to circular principles; reduction in virgin material throughput; waste elimination rates	5–15 years	Material flow analysis; sector surveys
Natural capital stock change	Changes in forest cover, soil quality, water availability, biodiversity indices, and coastal ecosystem health in ecosystem territory	10–30 years	Natural capital accounts; ecosystem service assessments
Environmental governance transformation	Changes in territorial environmental policy frameworks, planning standards, and institutional practices attributable to ecosystem influence	5–15 years	Policy document analysis; institutional surveys
Climate resilience capacity	Territory's demonstrated capacity to absorb and recover from climate-related shocks without loss of social or environmental function	10–25 years	Resilience assessments; disaster recovery data
Environmental knowledge institutionalization	Degree to which ecosystem-generated environmental knowledge has been embedded in territorial planning, regulatory, and educational systems	5–15 years	Knowledge uptake tracking; institutional analysis

Source Authors' own elaboration

6.2.2 *Social Transformation*

The long-term social impact of a purpose ecosystem is assessed at the level of structural changes in the conditions of social life in a territory, namely changes in the distribution of capabilities and opportunities, changes in the quality and density of social relationships, changes in the cultural norms and institutional practices that determine what is possible for different communities, and changes in the political voice and agency of historically marginalized groups. These are changes that occur over years and decades through processes that are cumulative, non-linear, and deeply embedded in the specific social histories of particular territories. They cannot be produced by any single social service program, however well-designed. They require the sustained, coordinated engagement of multiple actors around a shared commitment to social transformation.

Sen's (1999) capabilities approach, discussed in depth in Chap. 3, provides the most theoretically rigorous framework for assessing long-term social impact. Impact, in the capabilities framework, means the expansion of the substantive freedoms that people enjoy, their real opportunities to live long and healthy lives, to be educated, to participate in political life, to engage in meaningful work, and to maintain the social relationships that sustain human flourishing. Long-term social impact assessment asks whether these capabilities have genuinely expanded for the communities the ecosystem serves, particularly for those who were most constrained at the outset. It asks not merely whether services have been provided or programs delivered, but whether the conditions that determine what is possible for people have changed in durable and meaningful ways.

Fraser's (1995, 2013) three-dimensional justice framework, also discussed in Chap. 3, provides a complementary evaluative lens that ensures long-term social impact assessment attends to all the dimensions of social transformation rather than only the most easily measured economic ones. Redistribution (the reduction of economic inequality through changes in the structures that generate it) requires evidence of long-term changes in income distribution, asset ownership, and economic opportunity. Recognition (the transformation of cultural norms that devalue certain identities and groups) requires evidence of changes in how communities understand and relate to diversity, marginalized identities, and minority cultures. Representation (the expansion of political voice and institutional access for historically excluded groups) requires evidence of changes in governance structures, political participation, and institutional responsiveness to diverse community needs. All three dimensions are necessary for genuine long-term social transformation, and impact assessment frameworks that attend only to economic redistribution will systematically miss the most important dimensions of change.

The social capital tradition of Putnam (1993, 2000) and Coleman (1990) provides another evaluative framework for long-term social impact, emphasizing the importance of the relational infrastructure of communities, namely the networks, norms of reciprocity, and trust relationships that enable collective action and generate community well-being. Long-term improvements in social capital (measured through indicators of civic engagement, associational density, intergroup trust, and community cohesion) represent some of the most important and most durable contributions that purpose ecosystems can make to the territories they serve. Communities with high social capital are more resilient in the face of adversity, more effective at coordinating collective responses to shared challenges, more capable of holding institutions accountable, and generate greater individual and collective well-being than communities with low social capital. Building social capital is a long-term project that cannot be accomplished through short-term interventions, but that represents exactly the kind of systemic change that purpose ecosystems are positioned to advance.

In Table 6.2, long-term impact indicators, related to social transformation, are presented.

6.2.3 Territorial Development

The long-term territorial development impact of a purpose ecosystem goes beyond the immediate economic effects of its activities (captured by the output and short-term outcome indicators of Chap. 3) to encompass the transformation of the territory's development trajectory, its capacity for innovation and learning, its attractiveness to aligned actors and resources, its institutional quality and governance effectiveness, and its distinctive identity and position within broader regional and national systems. Purpose ecosystems that generate genuine long-term territorial impact do not merely improve economic indicators in the short term. They change the kind of territory they are embedded in, building capabilities and institutions that sustain development long after specific ecosystem initiatives have concluded.

Territorial innovation systems research, associated with scholars such as Cooke (1992), Asheim and Gertler (2005), and Camagni (1991), provides the most developed framework for understanding how territorial configurations of actors, institutions, and relationships generate long-term development advantages. Innovative regions and territories are distinguished not primarily by the size of their economies or the scale of their public investments, but by the quality of their institutional environment for innovation, the density and quality of collaborative relationships among firms, universities, and public institutions; the effectiveness of knowledge transfer mechanisms; the availability of skilled human capital; the quality of entrepreneurial ecosystems; and the presence of cultural norms that support risk-taking, learning from failure, and collaborative problem-solving. Purpose ecosystems contribute to long-term territorial development by building and strengthening these institutional foundations for territorial innovation capacity.

Smart Specialization Strategies (S3), discussed in depth in Chap. 2, provide an important policy framework for understanding and assessing the territorial development impact of purpose ecosystems in European contexts. S3 aims to identify and develop the distinctive assets, capabilities, and knowledge bases of territories,

Table 6.2 Long-term impact indicators: social transformation

Long-term impact indicator	What it measures	Time horizon	Primary method
Capability expansion index	Changes in the substantive freedoms available to community members, measured across Sen's capability dimensions: Health, education, political participation, economic opportunity, social affiliation	10–25 years	Longitudinal capability surveys; administrative data
Structural inequality reduction	Long-term changes in the distribution of income, assets, educational attainment, and economic opportunity, with particular attention to historically excluded groups	10–20 years	Longitudinal income and wealth data; administrative records
Social capital development	Changes in civic engagement rates, associational density, inter-group trust, community cohesion, and collective efficacy in the ecosystem territory	5–15 years	Longitudinal social capital surveys; civic participation data
Cultural recognition and vitality	Changes in the cultural recognition and participation of minority, indigenous, and marginalized communities; vitality of cultural institutions and practices	10–20 years	Cultural participation surveys; language vitality indicators
Political representation and voice	Changes in the representation of historically excluded groups in governance bodies, political institutions, and civic leadership positions in the territory	5–15 years	Political representation data; governance surveys
Community agency and self-determination	Degree to which communities are driving their own development trajectories, making collective decisions about their futures, and holding institutions accountable	10–20 years	Participatory assessments; governance quality evaluations
Well-being and flourishing	Long-term changes in subjective Well-being, mental health, sense of belonging, and quality of social relationships among ecosystem-connected community members	5–15 years	Longitudinal Well-being surveys; health data

Source Authors' own elaboration

enabling them to develop competitive advantages in specific areas of economic and social innovation. Purpose ecosystems that are well-aligned with the S3 priorities of their territories can contribute to long-term S3 impact, strengthening the distinctive knowledge domains that the strategy targets, attracting external investment and talent to the territory, and building the collaborative relationships among universities, firms, and public institutions that are the foundation of competitive territorial innovation systems. Assessing purpose ecosystem contribution to long-term S3 impact requires longitudinal analysis of territorial economic specialization, knowledge production, and innovation capacity, drawing on data sources such as patent records, publication databases, firm establishment and survival data, and regional economic statistics.

Territorial resilience is a concept developed at the crossroads of regional economics and complexity theory (Martin, 2012; Simmie and Martin, 2010). This concept is useful to understand how purpose ecosystems strengthen the capacity of a territory to adapt to continuous new conditions. Territorial resilience is not simply relating to facing economic shocks. It is about the ability of a territory to reshape its economy, institutions, and social capabilities, always based on its core assets, relationships, and values, in response to shifting environmental, technological, or economic conditions. Purpose ecosystems contribute to this ability by broadening the economic base of their territories, by building the social trust and civic capacity needed to face adversity together, by developing institutions that can adapt over time, and by protecting the environmental assets that underpin the long-term health of the territory.

Table 6.3 shows long-term impact indicators in relation to territorial development dimension.

6.2.4 *Ecosystem Institutional Legacy*

The fourth dimension of long-term ecosystem impact is perhaps the most theoretically distinctive and the most difficult to measure. It is the institutional legacy that a purpose ecosystem leaves in its territory. Institutional legacy refers to the changes in formal and informal institutions, organizational forms, cultural norms, and governance practices that persist after specific ecosystem initiatives have concluded, that were produced through or significantly shaped by ecosystem activities, and that continue to generate social and environmental value long after the ecosystem itself may have evolved or dissolved. It is the most direct expression of what it means for an ecosystem to have genuinely transformed the conditions, rather than merely the outcomes, of its territory.

New institutionalism in organizational theory, associated with scholars such as DiMaggio and Powell (1983), Scott (1995), and North (1990), provides the theoretical foundation for understanding institutional change as a dimension of ecosystem impact. Institutions, in this tradition, encompass not only formal rules and organizations but also the informal norms, cognitive frameworks, and cultural practices that structure social interaction and shape what actors perceive as possible, desirable, and legitimate. Institutional change is the most durable and most consequential form of social change, because institutions determine the rules of the game within which all other social activities occur. A purpose ecosystem that succeeds in changing the institutional rules of its territory (whether by establishing new governance norms, creating new organizational forms that become templates for future organizations,

Table 6.3 Long-term impact indicators: Territorial development

Long-term impact indicator	What it measures	Time horizon	Primary method
Territorial innovation capacity	Long-term changes in the territory's capacity for knowledge production, innovation, and learning, measured through patent activity, research output, new firm formation, and knowledge worker density	10–20 years	Patent data; publication databases; firm registers
Smart specialization advancement	Progress in the development of the territory's distinctive knowledge domains and competitive advantages as defined by S3 priorities	5–15 years	Sector analysis; investment tracking; S3 evaluation data
Talent and investment attraction	Long-term changes in the territory's capacity to attract and retain skilled workers, researchers, and aligned investment	5–15 years	Migration data; investment statistics; graduate retention rates
Territorial economic resilience	Demonstrated capacity to adapt economic structure and maintain employment and Well-being through major shocks	10–25 years	Economic shock analysis; recovery trajectory data
Institutional quality improvement	Long-term changes in the quality, responsiveness, and effectiveness of territorial governance and public institutions	10–20 years	Institutional quality surveys; governance assessments
Cross-sectoral collaboration density	Permanence and depth of collaborative relationships among firms, universities, and public institutions in the territory	5–15 years	Network analysis; collaboration surveys
Territorial identity and reputation	Development of a distinctive territorial identity associated with transformative purpose that attracts aligned actors and resources	10–20 years	Reputation surveys; media analysis; actor attraction data

Source Authors' own elaboration

embedding new values in professional practices, or shifting the cultural expectations of what economic and civic actors owe to each other and to their shared environment) has achieved a form of impact that will outlast any specific program, project, or funding cycle.

Systemic change theory, developed in the social innovation literature by Westley et al. (2011) and Mulgan (2019), provides a complementary framework for understanding ecosystem institutional legacy. Genuine systemic change occurs when innovations in solutions, strategies, and social relationships become embedded in the rules, norms, and incentive systems that govern a social domain. The most important indicators of systemic change are therefore not the outcomes of specific initiatives but the changes in the systems that produce those outcomes, namely changes in the policies that govern a sector, changes in the professional norms that guide practice, changes in the organizational forms that deliver services, and changes in the public narratives that define what is considered normal, desirable, and legitimate. Purpose ecosystems that generate genuine systemic change leave territories with different institutional DNA than they found. They are more collaborative, more oriented toward social and environmental purpose, more capable of sustained collective action, and more resistant to the concentration of power and resources in the hands of actors whose interests diverge from community well-being.

Platform legacy, drawn from the platform strategy literature (Gawer & Cusumano, 2014; Parker et al., 2016), offers another perspective on ecosystem institutional legacy that is particularly relevant in contexts where digital infrastructure has played an important role in ecosystem development. Purpose ecosystems that develop shared digital platforms, data systems, and governance architectures create technological infrastructure that can outlast the ecosystem itself and continue to generate value for the territory long after the ecosystem's formal activities have concluded. Open-source software, shared data standards, interoperable data systems, and community-owned digital platforms are examples of technological institutional legacy that embeds ecosystem values in the digital infrastructure of a territory. Assessing this dimension of institutional legacy requires attention not only to the social and organizational changes the ecosystem has generated but to the technological architectures and governance systems it has built.

In Table 6.4, long-term impact indicators of ecosystem institutional legacy are presented.

6.3 Attribution, Governance, and Legitimacy in Impact Assessment Practice

Framework design and indicator selection are only the beginning of the challenge. The most demanding questions in ecosystem impact assessment are practical: How do we demonstrate that observed systemic changes are meaningfully connected to ecosystem activities rather than to other factors? How do impact assessment findings feed into the governance and collective learning of purpose ecosystems? And how does the practice of impact measurement contribute to, or undermine, the legitimacy that purpose ecosystems depend on?

They are also the questions that most clearly distinguish impact assessment in purpose ecosystems from conventional organizational performance management. In conventional organizations, performance measurement is primarily an internal management tool. Data is collected to inform strategic decisions, allocate resources, and hold managers accountable to organizational objectives. The relationship between measurement and governance is relatively straightforward. In purpose ecosystems, this simple logic breaks down. Impact does not belong to any single actor and cannot be directed by any single governance structure. The relationship between

Table 6.4 Long-term impact indicators: Ecosystem institutional legacy

Long-term impact indicator	What it measures	Time horizon	Primary method
Policy and regulatory transformation	Changes in territorial and national policy frameworks, regulatory standards, and public investment priorities that can be traced to ecosystem influence and advocacy	5–15 years	Policy document analysis; stakeholder interviews; legislative tracking
New organizational forms and models	Emergence and diffusion of new organizational forms, business models, and governance structures generated or inspired by ecosystem activities	5–15 years	Organizational registries; diffusion tracking; case studies
Professional norm transformation	Changes in the professional norms, practices, and standards of the sectors in which the ecosystem operates, attributable to ecosystem influence	10–20 years	Professional surveys; sector analyses; comparative studies
Governance innovation diffusion	Degree to which governance innovations developed within the ecosystem have been adopted by other organizations and institutions in the territory and beyond	5–15 years	Adoption tracking; comparative governance analysis
Cultural norm change	Long-term shifts in the cultural norms, public narratives, and social expectations that govern economic and civic behavior in the territory	15–30 years	Longitudinal cultural surveys; discourse analysis; ethnographic studies
Institutional capacity building	Permanent expansion of the institutional capacity of community organizations, civic institutions, and public bodies to pursue transformative purpose independently	10–20 years	Organizational capacity surveys; institutional longitudinal studies
Shared infrastructure legacy	Continued generation of value from shared platforms, data systems, knowledge bases, and physical infrastructure developed through ecosystem activities	5–20 years	Platform usage tracking; infrastructure value assessment

Source Authors' own elaboration

measurement, governance, and legitimacy is more complex, more distributed, and more dependent on the quality of the collaborative relationships through which measurement is designed, conducted, and interpreted.

The attribution challenge is the most technically demanding dimension of ecosystem impact assessment and the one that most clearly differentiates it from conventional organizational evaluation. Attribution, in evaluation methodology, refers to the causal claim that observed outcomes were produced by a specific intervention or actor rather than by other factors. In simple, controlled settings, attribution can be established through randomized controlled trials, quasi-experimental designs, and other approaches that create comparison groups against which the effect of an intervention can be estimated. In complex, multi-actor, long-term systems like purpose ecosystems, these methods are rarely applicable or appropriate. The ecosystems cannot be randomized. The relevant outcomes emerge over decades, not months. The causal mechanisms involve the interaction of multiple factors that cannot be controlled or isolated. And the most important dimensions of impact (the institutional and cultural changes that represent genuine systemic transformation) resist quantitative measurement altogether.

Contribution analysis, as developed by Mayne (2001) and refined by Befani and Mayne (2014), provides the most rigorous and practically applicable methodology for addressing the attribution challenge in complex systems. The core insight is that the goal of impact evaluation in complex settings should not be to prove causation (which is rarely achievable) but to demonstrate that specific activities made a meaningful and identifiable contribution to observed outcomes, through mechanisms that can be articulated and tested against evidence, in ways that provide a reasonable basis for inferring that the outcomes would have been less likely, less complete, or slower to emerge without those activities. This is a more modest but also more honest standard of evidence than causal attribution, and it is much better suited to the reality of complex, multi-causal systems.

The contribution analysis methodology involves several interconnected steps. A narrative account of how ecosystem activities are expected to contribute to observed systemic changes, specifying the causal mechanisms involved, the contextual conditions that must be present for those mechanisms to operate, and the alternative explanations that might account for observed changes without ecosystem contribution. Besides, data are collected on whether the proposed mechanisms are operating as expected, whether the contextual conditions are present, and whether the alternative explanations can be ruled out or rendered less plausible. The contribution story is also revised in light of the evidence, strengthening the elements supported by data, qualifying those that are uncertain, and acknowledging the limits of what can be claimed. At the end, it is validated with key stakeholders to assess its plausibility from multiple perspectives.

Process tracing, as developed in political science and applied to evaluation by Beach and Pedersen (2013) and Befani and Stedman-Bryce (2017), provides a complementary methodology that is particularly valuable for assessing ecosystem contribution to institutional and governance changes. Process tracing examines the within-case causal mechanisms through which ecosystem activities are connected to observed outcomes, asking not merely whether correlation exists between ecosystem activities and outcomes but whether the specific causal mechanisms proposed in the contribution story can be observed operating in the specific case. Evidence of mechanism operation (such as documents showing that ecosystem-generated knowledge influenced a specific policy decision, or interviews revealing that governance practices adopted by a public institution were inspired by ecosystem governance innovations) provides much stronger support for contribution claims than correlational evidence alone.

Participatory evaluation approaches (in which community members and ecosystem actors are involved as active co-producers of impact knowledge rather than passive subjects of external assessment) address a different limitation of conventional impact evaluation, its tendency to privilege the knowledge and perspectives of professional evaluators over the lived experience and contextual knowledge of those most directly affected by ecosystem activities. Communities often have access to forms of knowledge about social and environmental change in their territories that are invisible to external evaluators. They notice when relationships change, when institutions become more responsive, when cultural norms shift, when the territory feels different. This knowledge is not always expressible in the quantitative terms that conventional evaluation frameworks privilege, but it is often the most meaningful evidence available about whether genuine transformation is occurring. Participatory evaluation creates spaces for this knowledge to be shared, documented, and integrated with more formal evidence, producing impact assessments that are richer, more credible, and more legitimate in the eyes of the communities they concern.

The relationship between impact assessment and ecosystem governance is the second major dimension of the practical challenge addressed in this section. In emergent and policy-initiated ecosystems, impact assessment is most valuable when it functions not merely as a reporting mechanism but as a generator of collective learning that informs governance decisions and adaptive management. This requires creating explicit links between the impact assessment process and the governance processes through which ecosystem actors make collective decisions about strategy, resource allocation, and organizational adaptation.

In organically emergent ecosystems, these links are necessarily informal. Impact evidence circulates through the ecosystem in the form of stories, narratives, and informal data sharing rather than formal reports. When an ecosystem actor observes that a particular approach is generating strong social transformation results, or that an environmental intervention is not producing the expected systemic change, this information is shared through the relational networks of the ecosystem, influencing the decisions and activities of other actors in informal but real ways. The quality of these informal learning processes depends on the density and quality of the relational fabric of the ecosystem. Ecosystems with rich, trusting, and honest relationships among their actors are better able to share impact evidence and learn collectively from it than ecosystems characterized by competition, information hoarding, or institutional defensiveness.

In policy-initiated ecosystems, the links between impact assessment and governance can and should be more formal. Regular evaluation cycles (in which accumulated impact data are reviewed by governance bodies and used to inform strategic decisions) create an institutional basis for adaptive management. This requires designing the governance structure of policy-initiated ecosystems to include explicit evaluation and learning functions, namely regular review processes in which impact data are presented and discussed, mechanisms for translating evaluation findings into governance decisions, and institutional cultures that welcome critical feedback and are genuinely willing to change course when evidence demands it. The adaptive governance literature, discussed in Chap. 5, provides guidance on how to design governance systems with these learning capabilities, emphasizing the importance of feedback loops, polycentricity, and institutional flexibility as design principles for adaptive governance in complex systems (Folke et al., 2005; Ostrom, 2010).

The risk of indicator fixation deserves specific attention in the context of ecosystem governance. Indicator fixation, described by Power (1999) as the audit society's tendency to privilege what can be measured over what matters, is a persistent risk in any performance management system. When governance decisions are driven primarily by indicator performance, organizations and ecosystems develop incentives to optimize for measurable outcomes at the expense of the deeper, less measurable changes that represent genuine transformative impact. This risk is particularly acute for purpose ecosystems, because the most important dimensions of long-term impact (institutional and cultural transformation) are precisely those that are most resistant to quantitative measurement. Governance processes that privilege easily quantified short-term metrics will systematically undervalue the most significant work that purpose ecosystems do. The remedy is not to abandon measurement but to design governance processes that combine quantitative impact data with rich qualitative evidence, community narratives, and honest acknowledgment of what cannot be measured without losing its meaning.

Legitimacy is another dimension of the practical challenge addressed in this section, and the one that most directly connects impact assessment to the purpose ecosystem's relationship with the communities it serves. Legitimacy, in the organizational theory tradition associated with Suchman (1995) and Scott (1995), is the generalized perception or assumption that an organization's actions are desirable, proper, and appropriate within some socially constructed system of norms, values, beliefs, and definitions. For purpose ecosystems, legitimacy is not primarily a reputational asset to be managed but a relational quality that must be continuously earned through patterns of behavior that demonstrate genuine commitment to Transformative Purpose.

Impact assessment contributes to legitimacy when it is honest, transparent, and genuinely responsive to community perspectives. An ecosystem that conducts rigorous impact assessment, shares findings openly (including those that reveal failures or disappointing results) and uses impact evidence to improve its activities rather than merely to communicate successes, demonstrates a form of accountability that builds deep legitimacy with the communities it serves. This is the legitimacy that comes from being trustworthy, not merely from looking good. It is the only form of legitimacy that is durable in the face of scrutiny and that can sustain the long-term collaborative relationships on which purpose ecosystem effectiveness depends.

The risk of purpose-washing is directly connected to the quality of impact assessment practices. Purpose-washing occurs when ecosystems adopt the language of transformative commitment without the substance, when they report on easily measured outputs while avoiding honest assessment of whether systemic impact is being generated, when they attribute favorable environmental and social trends to ecosystem activities without credible contribution analysis, or when they use impact communications primarily as fundraising and reputation management tools rather than as accountability mechanisms. Rigorous, honest, and participatory impact assessment is one of the most important structural safeguards against purpose-washing, because it creates the institutional infrastructure and the relational norms that make it difficult to maintain the gap between rhetoric and reality over time.

Impact assessment also contributes to legitimacy by demonstrating that the ecosystem is a responsible steward of public and private resources. For policy-initiated ecosystems that receive significant public funding, impact assessment provides the accountability evidence that justifies continued public investment. For ecosystems that seek private and philanthropic investment, impact assessment provides the credibility evidence that enables capital mobilization. For ecosystems that depend on the voluntary contribution of community members and civic organizations, impact assessment provides the evidence of genuine progress that sustains motivation and commitment over the long haul. In all these cases, the quality of impact assessment (whether rigorous, honest, and community-centered or superficial, selective, and institution-serving) is a direct expression of the ecosystem's values and a key determinant of its long-term viability.

Purpose-washing prevention has a structural dimension that deserves emphasis. Purpose-washing is not always intentional. Ecosystems can drift toward rhetorical purpose without substantive commitment through gradual processes of institutional pressure, resource competition, and the accumulation of small compromises that individually seem reasonable but collectively erode genuine purpose commitment. Regular, rigorous, and community-anchored impact assessment creates institutional resistance to this drift by making the gap between rhetoric and reality visible before it becomes irreversible. When impact assessment findings consistently show that the ecosystem is not generating the social and environmental changes it claims, that gap cannot be sustained indefinitely in the face of honest community engagement. This is why impact assessment is not merely a reporting mechanism but a governance function: it is one of the primary ways in which purpose ecosystems maintain accountability to the transformative commitments that give them their identity and legitimacy.

Several practical recommendations guide the development of legitimate, learning-oriented impact assessment systems in purpose ecosystems. Naturally, no evaluation framework can definitively resolve the tension between attribution and contribution. The proposed model should therefore be understood as a guiding tool, rather than as a device for certifying impact.

Before selecting indicators, we need to develop a clear and explicit theory of change that articulates how ecosystem activities are expected to contribute to systemic impact through specific mechanisms. Indicators without theory produce data without meaning.

Moreover, the communities whose transformation is being assessed should be active participants in defining what impact means, what indicators are meaningful, and how evidence should be interpreted.

Investing in longitudinal data infrastructure is a long-term investment that pays dividends in the quality of impact evidence available. Long-term impact is invisible in cross-sectional snapshots, and we need to build data systems that track territorial conditions over years and decades.

In addition, quantitative indicators provide breadth; qualitative case studies provide depth; participatory evaluation provides legitimacy. All three are necessary for credible impact assessment in complex systems. Be honest about what you do not know. The limits of attribution in complex systems are real, and pretending otherwise undermines credibility.

Finally, impact assessment findings should generate governance responses, not just reports. Building explicit mechanisms for translating impact evidence into collective decisions is the most important step in transforming impact assessment from a compliance burden into a generative practice.

Box 6.2 A Framework for Impact-Informed Governance

Step 1: Theory of change development. Before collecting data, articulate explicitly how ecosystem activities are expected to contribute to systemic change through specific mechanisms. Update the theory of change regularly as the ecosystem learns.

Step 2: Participatory indicator selection. Involve community members, ecosystem actors, and external stakeholders in selecting the indicators that matter most for assessing progress toward Transformative Purpose in the specific context.

Step 3: Longitudinal data collection. Build data systems that track territorial conditions over years and decades. Combine administrative data with periodic surveys, network analyses, and qualitative case studies.

Step 4: Contribution analysis. Develop and test contribution stories that connect ecosystem activities to observed systemic changes through identifiable mechanisms, acknowledging alternative explanations and the limits of attribution.

Step 5: Community validation. Share impact assessment findings with community members and stakeholders. Invite challenge, correction, and supplementation from those with direct experience of the changes being assessed.

Step 6: Governance response. Bring validated impact findings into governance processes. Use them to inform strategic decisions, reallocate resources, and adapt ecosystem activities. Document the governance decisions that impact evidence informed.

Step 7: Public accountability. Share impact assessment findings transparently with funders, policymakers, and the public. Include honest reporting on failures and disappointments as well as successes. Use public reporting as a demonstration of accountability rather than a communications exercise.

Spotlight 6.2 Impact Assessment in the EU Mission on Climate-Neutral and Smart Cities

In 2022, the European Commission selected 100 European cities to participate in the EU Mission on Climate-Neutral and Smart Cities, one of five Horizon Europe missions designed to tackle major societal challenges through coordinated multi-actor innovation. The cities' mission was explicit and time-bound: achieve climate neutrality by 2030, two decades ahead of the EU's 2050 climate target. The 100 cities committed to developing Climate City Contracts, so formal multi-stakeholder agreements between the city, its citizens, businesses, and national government specifying the transformation pathways through which each city would achieve its target. The contracts were co-developed with community participation. In most cities, the process involved extensive civic engagement through workshops and deliberative processes that brought residents, businesses, and civic organizations into the definition of transformation goals.

The Climate City Contract is, from the perspective of this chapter, one of the most ambitious impact assessment instruments ever developed for a policy-initiated purpose ecosystem. Each contract is a governance document and an impact framework. It specifies the baseline conditions from which the city is starting, the transformation pathways it has committed to pursue, the intermediate milestones against which progress will be assessed, and the mechanisms through which the city will maintain accountability to its citizens and the European Commission simultaneously. This dual accountability structure (downward to citizens and upward to the Commission) operationalizes the accountability architecture that this chapter argues is essential for genuine purpose ecosystem governance.

What makes the Mission on Climate-Neutral Cities particularly instructive is the sophistication of its approach to the attribution problem. The Mission's impact framework explicitly acknowledges that climate neutrality in a city cannot be attributed to any single actor, program, or policy. Urban carbon emissions are the aggregate result of decisions made by millions of households, businesses, and public institutions. What the Climate City Contracts measure is not the isolated contribution of Mission-funded activities but the trajectory of the city as a whole: whether total emissions are declining at the rate required to reach climate neutrality by 2030, and whether the governance arrangements, policy changes, and cultural shifts necessary for that trajectory are developing as planned. The Mission distinguishes explicitly between output, outcome, and systemic impact levels. Renewable energy installations are outputs, changes in energy consumption patterns are outcomes, and the transformation of planning regulations and cultural norms is the systemic impact.

The 100 Mission cities also function as a learning ecosystem for impact assessment. Each city's experience with measuring climate impact, tracking governance changes, and assessing community engagement generates knowledge shared across the network through the NetZeroCities platform, enabling collective learning about what works in the transition to urban climate neutrality. This meta-level learning function (through which the Mission as a whole learns from the diverse experiences of its 100 participating cities) is itself an important impact pathway: governance innovations and measurement approaches that prove effective in one city can be adapted and applied in others, multiplying the impact of each successful experiment across the entire network. The Mission on Climate-Neutral Cities already represents one of the most sophisticated available operationalizations of the contribution-focused, systemic, longitudinal impact framework that this chapter argues all-purpose ecosystems need to develop.

6.4 A Direction, Not a Destination...

A compass helps you understand where you are and which direction you are facing, and what lies ahead if you hold your course. It does not tell you where to go. The Purpose Ecosystem Compass was built in that spirit, not as a blueprint but as an instrument of orientation. Purpose ecosystems are organizational forms through which communities and territories commit, collectively and over time, to moving toward a social and environmental purpose. This is not reachable in any final sense, but the direction of travel matters and traveling together makes the journey possible at all.

Markets have proven insufficient to address the structural roots of inequality and ecological breakdown. Governments struggle to sustain the long-term commitments that genuine transformation demands. A growing body of scholarship has documented this inadequacy and the search for new forms of collective action adequate to it, from Mazzucato's (2018) work on mission-oriented innovation to Sachs et al.'s (2019) frameworks for sustainable development governance, from Mulgan's (2019) account of social innovation to Kania and Kramer's (2011) collective impact tradition. What this book has tried to add to that conversation is a framework for understanding what makes them work and what makes them fail.

That framework rests, above all else, on a distinction between authentic purpose and purpose as legitimating language. The problem of purpose-washing, described by Crane et al. (2014), has grown sharper with every passing year as social and environmental commitments have become more rhetorically expected and more structurally difficult to maintain. The ecosystems that genuinely advance transformation are those in which direction governs decisions when resources are scarce, interests diverge, and the pressure to prioritize institutional survival over mission becomes acute. And the central argument of this book is that the four components of the PEC, enabling conditions, Transformative Purpose, digital mechanisms, and governance architecture, matter precisely because each of them, in its own way, determines whether authentic purpose can emerge, survive, and deepen over time, or whether it gradually dissolves into purposelessness.

For policymakers, this has implications that go beyond the design of any particular program or funding instrument. The past decade has seen substantial public investment in ecosystem-building initiatives, from Italy's PNRR-funded national innovation ecosystems to the many territorial development programs that have sought to build collaborative multi-actor structures at regional and national scale. These investments are not misguided. Enabling conditions matter, and building them is a legitimate and important task for public institutions. What the PEPI framework asks, however, is whether the territories they are investing in are ready for the purpose that institutional structure is supposed to serve, whether the social capital, institutional quality, and relational trust are already present, or whether they require prior cultivation. Ecosystem policy that bypasses this question, and invests in containers without attending to content, is building compasses for people who do not yet share a direction. The purpose an ecosystem is built to serve cannot be prescribed from above, only cultivated from within, and the most important thing a policymaker can do is create the conditions in which that cultivation becomes possible (Ansell & Gash, 2008; Ostrom, 2010).

In this sense, the relationship between the Purpose Ecosystem Compass and impact is constitutive rather than instrumental. When all four components are present and genuinely aligned, the ecosystem becomes the kind of collective actor for which long-term systemic change is a realistic aspiration. The four long-term impact dimensions, environmental sustainability transitions, social transformation, territorial development, and ecosystem institutional legacy, are not metrics imposed on this alignment from outside. They are the forms in which a sustained, aligned, and authentically purposeful ecosystem eventually becomes visible in a territory (Emerson & Nabatchi, 2015; Patton, 2011).

For communities, the message is clear. The emergence of purpose ecosystems across many territories is a story about communities. Mondragon grew from a community's decision, in the aftermath of civil war, to build its own future rather than wait for someone else to build it. Bologna's collaborative governance emerged from civic actors who recognized that shared spaces could only be genuinely cared for by the people who shared them. Preston's community wealth building was invented by a council and its partners who refused to accept that their direction of travel was fixed by decisions made elsewhere.

For scholars, the framework raises questions concerning the dynamics of purpose itself, how it emerges in specific institutional and territorial conditions, how it survives or fails when ecosystems face resource pressures or leadership transitions, and what drives its erosion over time (Battilana & Dorado, 2010; George et al., 2023). Enabling conditions are not neutral. Particular configurations of social capital, institutional quality, and civic infrastructure make certain forms of Transformative Purpose possible and others difficult to sustain. Understanding that relationship more precisely would tell us a great deal about why purpose ecosystems take different shapes in different territories, and what systemic change can look like in each of them. Other questions are related to governance. Redistributing power toward community is easy to endorse but difficult to do, especially in ecosystems where resource and institutional asymmetries are significant. Future studies could follow ecosystems over time to understand which concrete practices power shifts toward communities. The question also remains open of how what we have called structural reflexivity is sustained over time. This is the practice of continually questioning whether governance structures are still serving their purpose, or are slowly stifling it. Future research could explore which institutional mechanisms help keep this practice alive. In addition, the paradox of structure is a tension to be managed. For this reason, comparative studies on ecosystems that have gone through governance crises would offer a valuable contribution, as they would make it possible to empirically observe how that tension is managed. Moreover, governance under conditions of strong power asymmetry between actors remains to be further explored. This is a condition as common in practice as it is neglected in theory. Here what is needed is empirical research on concrete cases in which actors highly unequal in resources and institutional legitimacy have had to share the same decision-making table, in order to understand which governance arrangements protected the voice of weaker actors, and which instead only formally represented it. Finally, the question of measurement. How do we assess what purpose ecosystems contribute to long-term systemic change in ways that are honest about causality and useful to the communities themselves? Future research can adopt adequate approaches with longitudinal and participatory research designs.

Purpose ecosystems are, in the end, a collective act of orientation, the ongoing and relational work of communities and territories deciding together who they want to become. The destination may always be ahead. The direction is something we choose, together, now.

References

Ansell, C., & Gash, A. (2008). Collaborative governance in theory and practice. *Journal of Public Administration Research and Theory, 18*(4), 543–571.

Asheim, B. T., & Gertler, M. S. (2005). The geography of innovation: Regional innovation systems. In J. Fagerberg, D. C. Mowery, & R. R. Nelson (Eds.), *The Oxford handbook of innovation* (pp. 291–317). Oxford University Press.

Battilana, J., & Dorado, S. (2010). Building sustainable hybrid organizations: The case of commercial microfinance organizations. *Academy of Management Journal, 53*(6), 1419–1440.

Beach, D., & Pedersen, R. B. (2013). *Process-tracing methods: Foundations and guidelines*. University of Michigan Press.

Befani, B., & Mayne, J. (2014). Process tracing and contribution analysis: A combined approach to generative causal inference for impact evaluation. *IDS Bulletin, 45*(6), 17–36.

Befani, B., & Stedman-Bryce, G. (2017). Process tracing and Bayesian updating for impact evaluation. *Evaluation, 23*(1), 42–60.

Camagni, R. (1991). *Innovation networks: Spatial perspectives*. Belhaven-Pinter.

Coleman, J. S. (1990). *Foundations of social theory*. Harvard University Press.

Cooke, P. (1992). Regional innovation systems: Competitive regulation in the new Europe. *Geoforum, 23*(3), 365–382.

Corrente, S., Greco, S., Nicotra, M., Romano, M., & Schillaci, C. (2019). Evaluating and comparing entrepreneurial ecosystems using SMAA and SMAA-S. *The Journal of Technology Transfer, 44*(2), 485–519.

Costanza, R., d'Arge, R., de Groot, R., Farber, S., Grasso, M., Hannon, B., Limburg, K., Naeem, S., O'Neill, R. V., Paruelo, J., Raskin, R. G., Sutton, P., & van den Belt, M. (1997). The value of the world's ecosystem services and natural capital. *Nature, 387*(6630), 253–260.

Costanza, R., de Groot, R., Sutton, P., van der Ploeg, S., Anderson, S. J., Kubiszewski, I., Farber, S., & Turner, R. K. (2014). Changes in the global value of ecosystem services. *Global Environmental Change, 26*, 152–158.

Crane, A., Palazzo, G., Spence, L. J., & Matten, D. (2014). Contesting the value of "creating shared value". *California Management Review, 56*(2), 130–153.

DiMaggio, P. J., & Powell, W. W. (1983). The iron cage revisited: Institutional isomorphism and collective rationality in organizational fields. *American Sociological Review, 48*(2), 147–160.
Eisenhardt, K. M., & Graebner, M. E. (2007). Theory building from cases: Opportunities and challenges. *Academy of Management Journal, 50*(1), 25–32.
Emerson, K., & Nabatchi, T. (2015). *Collaborative governance regimes*. Georgetown University Press.
Espeland, W. N., & Sauder, M. (2007). Rankings and reactivity: How public measures recreate social worlds. *American Journal of Sociology, 113*(1), 1–40.
Ebrahim, A. (2019). *Measuring social change: Performance and accountability in a complex world*. Stanford University Press.
Ebrahim, A., & Rangan, V. K. (2014). What impact? A framework for measuring the scale and scope of social performance. *California Management Review, 56*(3), 118–141.
Fals Borda, O. (1987). The application of participatory action research in Latin America. *International Sociology, 2*(4), 329–347.
Folke, C., Hahn, T., Olsson, P., & Norberg, J. (2005). Adaptive governance of social-ecological systems. *Annual Review of Environment and Resources, 30*, 441–473.
Fraser, N. (1995). From redistribution to recognition? Dilemmas of justice in a post-socialist age. *New Left Review, 212*, 68–93.
Fraser, N. (2013). *Fortunes of feminism: From state-managed capitalism to neoliberal crisis*. Verso.
Gawer, A., & Cusumano, M. A. (2014). Industry platforms and ecosystem innovation. *Journal of Product Innovation Management, 31*(3), 417–433.
Geels, F. W. (2002). Technological transitions as evolutionary reconfiguration processes: A multi-level perspective and a case-study. *Research Policy, 31*(8–9), 1257–1274.
George, G., Haas, M. R., McGahan, A. M., Schillebeeckx, S. J. D., & Tracey, P. (2023). Purpose in the for-profit firm: A review and framework for management research. *Journal of Management, 49*(6), 1841–1869.
Kania, J., & Kramer, M. (2011). Collective impact. *Stanford Social Innovation Review, 9*(1), 36–41.
Krlev, G., Mildenberger, G., & Anheier, H. K. (2020). Innovation and societal transformation: What changes when the 'social' comes in? *International Review of Applied Economics, 34*(5), 529–540.
Martin, R. (2012). Regional economic resilience, hysteresis and recessionary shocks. *Journal of Economic Geography, 12*(1), 1–32.
Mayne, J. (2001). Addressing attribution through contribution analysis: Using performance measures sensibly *Canadian Journal of Program Evaluation, 16*(1), 1–24.
Mazzucato, M. (2018). *The value of everything: Making and taking in the global economy*. Allen Lane.
Morgan, S. L., & Winship, C. (2015). *Counterfactuals and causal inference: Methods and principles for social research* (2nd ed.). Cambridge University Press.
Mulgan, G. (2019). *Social innovation: How societies find the power to change*. Policy Press.
Nicotra, M. (2019). *Shifting paradigms in entrepreneurial ecosystems: A data-driven perspective*. UTET.
Nicotra, M., Romano, M., Del Giudice, M., & Schillaci, C. E. (2018). The causal relation between entrepreneurial ecosystem and productive entrepreneurship. *The Journal of Technology Transfer, 43*(3), 640–673.
North, D. C. (1990). *Institutions, institutional change and economic performance*. Cambridge University Press.
Ostrom, E. (2010). *Governing the commons: The evolution of institutions for collective action*. Cambridge University Press.
Parker, G. G., Van Alstyne, M. W., & Choudary, S. P. (2016). *Platform revolution*. Norton.
Patton, M. Q. (2011). *Developmental evaluation: Applying complexity concepts to enhance innovation and use*. Guilford Press.

Pel, B., Wittmayer, J. M., Dorland, J., & Jørgensen, M. S. (2020). Unpacking the social innovation ecosystem: An empirically grounded typology of empowering network constellations. *Innovation: The European Journal of Social Science Research, 33*(3), 311–336.

Power, M. (1999). *The audit society: Rituals of verification.* Oxford University Press.

Putnam, R. D. (1993). *Making democracy work: Civic traditions in modern Italy.* Princeton University Press.

Putnam, R. D. (2000). *Bowling alone: The collapse and revival of American community.* Simon and Schuster.

Ramalingam, B., Jones, H., Reba, T., & Young, J. (2008). Exploring the science of complexity: Ideas and implications for development and humanitarian efforts. *ODI Working Paper, 285.*

Reason, P., & Bradbury, H. (Eds.). (2008). *The SAGE handbook of action research: Participative inquiry and practice.* SAGE.

Rockstrom, J., Steffen, W., Noone, K., Persson, A., Chapin, F. S., Lambin, E. F., Lenton, T. M., Scheffer, M., Folke, C., Schellnhuber, H. J., Nykvist, B., de Wit, C. A., Hughes, T., van der Leeuw, S., Rodhe, H., Sorlin, S., Snyder, P. K., Costanza, R., Svedin, U., Falkenmark, M., Karlberg, L., Corell, R. W., Fabry, V. J., Hansen, J., Walker, B., Liverman, D., Richardson, K., Crutzen, P., & Foley, J. A. (2009). A safe operating space for humanity. *Nature, 461*(7263), 472–475.

Rogers, P. J. (2008). Using programme theory to evaluate complicated and complex aspects of interventions. *Evaluation, 14*(1), 29–48.

Romano, M., Cunningham, J., Cuttone, J., Munnia, A., & Nicotra, M. (2025). Intellectual capital-based research impact management: A strategic approach to increase regional intellectual capital. *Journal of Intellectual Capital, 26*(7), 24–42.

Romano, M., Nicotra, M., Corrente, S., & Castrogiovanni, A. (2020). Start-ups, innovation and performance: An empirical analysis using SMAA-S. In A. Caputo & M. Pellegrini (Eds.), *The entrepreneurial behaviour* (pp. 167–186). Emerald Publishing.

Sachs, J. D., Schmidt-Traub, G., Mazzucato, M., Messner, D., Nakicenovic, N., & Rockstrom, J. (2019). Six transformations to achieve the sustainable development goals. *Nature Sustainability, 2*(9), 805–814.

Scott, W. R. (1995). *Institutions and organizations.* SAGE.

Sen, A. (1999). *Development as freedom.* Oxford University Press.

Simmie, J., & Martin, R. (2010). The economic resilience of regions: Towards an evolutionary approach. *Cambridge Journal of Regions, Economy and Society, 3*(1), 27–43.

Suchman, M. C. (1995). Managing legitimacy: Strategic and institutional approaches. *Academy of Management Review, 20*(3), 571–610.

United Nations. (2021). *System of environmental-economic accounting: Ecosystem accounting (SEEA EA).* United Nations.

Westley, F., Zimmerman, B., & Patton, M. Q. (2011). *Getting to maybe: How the world is changed.* Vintage Canada.

Yin, R. K. (2014). *Case study research: Design and methods* (5th ed.). SAGE.

FSC
www.fsc.org
MIX
Papier | Fördert gute Waldnutzung
FSC® C083411

Kolibri 360 GmbH
Ferdinand-Jühlke-Straße 7
99095 Erfurt, Deutschland
produktsicherheit@kolibri360.de